A

GUIDE

TO

SCIENTIFIC AND TECHNICAL

JOURNALS

IN

TRANSLATION

(Second Edition)

Compiled by

Carl J. Himmelsbach
Ledgemont Laboratory
Kennecott Copper Corporation
Lexington, Massachusetts

and

Grace E. Brociner
Lincoln Laboratory
Massachusetts Institute of Technology
Lexington, Massachusetts

SPECIAL LIBRARIES ASSOCIATION
New York, New York
1972

© Special Libraries Association
235 Park Avenue South, New York, N.Y. 10003

ISBN 0-87111-213-2
LC 72-85955

Library of Congress Cataloging in Publication Data

Himmelsbach, Carl J
  A guide to scientific and technical journals in translation.

  1. Science—Periodicals—Translations into English—Bibliography. 2. Technology—Periodicals—Translations into English—Bibliography. I. Brociner, Grace E., joint author. II. Title.
Z7403.H5 1972   016.5'05   72-85955
ISBN 0-87111-213-2

Printed in the United States of America

PREFACE TO THE FIRST EDITION

During the past decade the number of scientific and technical journals appearing in translated format has grown at a furious pace. Added to this have been the many changes in original and translated titles and the separation of familiar publications into more specialized journals.

Unfortunately, while both the quantity and quality of translation journals have increased, there has been little coordination of these translations either in format or in extent of coverage. A veritable "translation jungle" has been the by-product of a haphazard approach to the problem.

It is the hope of the authors that this compilation will in some degree help to provide a modest guide through the "translation jungle." To achieve this goal, all journals (except those otherwise noted) have been examined in both the original and their translated counterparts. No attempt has been made to register an evaluation of the quality of the translations, as this is a very subjective matter. However, extent of coverage, status of volume table of contents and indices, extent of reference to original language pagination, relation of translation and original volume or issue numbers, changes in title of original or translation journals, "lag time," and publisher are provided for each publication.

In offering this compilation, the authors recognize the transitory nature of the information. Specific data for any item may become obsolete immediately after submission of this manuscript to the publisher. New translation journals may appear, or there may be changes in the title of the original language publication or its subdivision into more than one journal. It is hoped that a future edition of this monograph will be able to update any changes.

Finally, the authors would appreciate any comments, suggestions, or additional information which might make a future edition more useful and more accurate in any or all details.

PREFACE TO THE SECOND EDITION

Since the issuance of the first edition of this guide, little has changed in the way of lessening the twisted paths of the "translation jungle." Many more journals are now translated, some then translated have fallen by the wayside, and both original and translation publishers have changed titles, split journals, and even switched from serial to periodical issuance. The "translation jungle" continues to thrive.

On the bright side, we have found increased cooperation from the translation publishers, a greater tendency toward more accurate titles and relationships to the original language publications' volume or issue numbers and publication dates, and a more identifiable correlation of original and translation pages.

In addition to the many publishers who have cooperated in supplying information, we should also like to express our sincerest thanks to the many librarians who have made suggestions and afforded access to their collections to facilitate our research.

C.J.H.
G.E.B.

INTRODUCTION

Arrangement of all information presented is intended to provide quick and convenient reference to translated editions of original language journals. Since most references to the translation journals originate from citations to the original language publications, the order of presentation of data is arranged to meet this need:

    a. original language title;
    b. translated title;
    c. year and issue of original which first appears in translation;
    d. frequency (when known);
    e. publisher;
    f. additional data.

Each journal title will carry, in a simplified code, specific information intended to provide a key to the potential relevance of the translated publication to a user's need (which can very well range from a simple abstract to the necessity of consulting the complete text of an article).

Additional sections provide a list of addresses of all publishers, translated titles referenced to original language titles, some frequently encountered abbreviations of Russian language publications, and the Russian-English transliteration system used in this monograph.

Section II requires some explanation. A translation which does not gather its material from a specific original language publication is generally useless to anyone searching for translation of a citation. This is not intended to impugn the validity of such publications as scanners and alerters of foreign literature, but merely to forewarn potential users as to their limited retrospective retrieval value.

Section III cross-references translated titles to their original language counterparts. This section is provided primarily because so many translation publishers advertise their journals without adequate reference to the original language publication.

Since this compilation is restricted to translations of scientific and technical publications, the following categories have been excluded: political; economic (except as related to the economics of scientific and technical processes); social science; educational theory. It might be noted that most of these publications would fall within the category of Section II (as explained above).

Finally, it should be noted that all arrangement of information is intended for the use of librarians and scientific-technical retrospective searchers. Different arrangements might prove more useful for other purposes but they are not within the scope of this publication.

CONTENTS

Preface		i
Introduction		iii
PART I	Cover-to-Cover Translation Journals	1
PART II	Selections, Collections, and Other "Translation Journals"	29
PART III	Translated Titles Cross-Referenced to Original Titles	32
PART IV	Some Frequently Encountered Russian Abbreviations of Soviet Journals and Their Full Titles	39
PART V	Key to Publishers and Distributors Abbreviated in Individual Citations	44
PART VI	Russian Alphabet and Transliteration	49

PART I

COVER-TO-COVER TRANSLATION JOURNALS

EXPLANATION FOR PART I ENTRIES

** *Indicates that the title has been announced for publication in translation, but has not been so published as of the date of this study (June 1, 1972). Consult publisher for further information.*

Trans. begins — refers to year and volume of the original publication. Unless otherwise indicated, translated volume commences with issue number 1.

Trans. covers — refers to volume (or volumes) translated before translation ceased publication.

OP — following publisher code indicates back issues no longer available from the publisher.

Explanation of Coded data:

1. a — Cover-to-cover translation
   b — Selected articles translated

2. a — Full translations of all articles
   b — Some articles fully translated, others abridged or abstracted
   c — Selected articles abridged or abstracted
   d — Articles translated are fully translated, others merely cited or listed

3. a — Translation carries original language numbers
   b — Translation carries other numbers
   c — Translation carries no numbers

4. a — Translation carries original language dates
   b — Translation carries other dates
   c — Translation carries no dates

5. a — Translation identifies original pagination
   b — Translation does not identify original pagination

6. a — Translation has volume index or collective table of contents
   b — Translation has no volume index or collective table of contents

7. "Lag time" between month of issue of original publication and month of issue of translated counter-part (there is a possible variation of one month because of mailing practices both of the original and the translation)

8. Other information (specified)

9. Translation exists but has not been located for verification and evaluation

1 **ACTA BIOCHIMICA POLONICA**
Trans. begins with 1954, v. 1, continues through 1957, v. 4, begins again with 1961, v. 8+
Quarterly
Order from: NTIS
1.a; 2.a; 3.a; 4.a; 5.b; 6.a;
7. 6 mos.;
8. 1958–60, vols. 5–7 are in preparation.

2 **ACTA ENTOMOLOGICA SINICA**
Trans. covers 1965, v. 14.
Order from: ESA
9.

3 **ACTA MATHEMATICA SINICA**
(Chinese Mathematics)
Trans. covers 1962–1967, v. 1–9.
Order from: AMS
1.a; 2.a; 3.a; 4.a; 5.a; 6.a;
7. trans. suspended pending resumption of original.

4 **ACTA MECHANICA SINICA**
Trans. covers 1966, v. 9.
Quarterly
Order from: PLPC
1.a; 2.a; 3.a; 4.a; 5.a; 6.a;
7. trans. suspended pending resumption of original.

5 **ACTA MEDICA IUGOSLAVICA**
Trans. begins with 1962, v. 16+
3–4 issues per year
Order from: NTIS
1.a; 2.a; 3.a; 4.a; 5.a; 6.b;
7. 13 mos.

6 **ACTA PHYSICA SINICA**
(Chinese Journal of Physics)
Trans. covers 1965–1966, v. 21–22, nos. 1–8.
Order from: AIP
1.a; 2.a; 3.a; 4.a; 5.a; 6.a;
7. trans. suspended pending resumption of original;
8. there is also a "Chinese Journal of Physics," published in Taiwan in English for the Physical Society of the Republic of China.

7 **ACTA PHYSIOLOGICA POLONICA**
Trans. begins with 1962, v. 13+
Bimonthly
Order from: NTIS
1.a; 2.a; 3.a; 4.a; 5.a; 6.a;
7. 12 mos.

8 **ACTA POLONIAE PHARMACEUTICA**
Trans. begins with 1963, v. 20+
Bimonthly
Order from: NTIS
1.a; 2.a; 3.a; 4.a; 5.b; 6.a;
7. 10 mos.

AKADEMII NAUK DOKLADY see "Doklady Akademii Nauk SSSR"

AKADEMII NAUK SSSR IZVESTIYA see "Izvestiya Akademii Nauk SSSR"

9 **AKUSTICHESKII ZHURNAL**
(Soviet Physics—Acoustics)
Trans. begins with 1955, v. 1+
Quarterly
Order from: AIP
1.a; 2.a; 3.a; 4.b; 5.a; 6.a;
7. 4 mos.

10 **ALGEBRA I LOGIKA**
(Algebra and Logic)
Trans. begins with 1968, v. 7+
Bimonthly
Order from: PLPC
1.a; 2.a; 3.a; 4.a; 5.a; 6.a;
7. 12 mos.

11 **ANGEWANDTE CHEMIE**
(International Edition in English)
Trans. begins with 1962, v. 74+
Monthly
Order from: API
1.a; 2.a; 3.b; 4.a; 5.a; 6.a;
7. 2 weeks
8. 1962–1964, vols. 74–76, contain only selected articles.

12 **ANTIBIOTIKI**
(Antibiotics)
Trans. covers 1959, v. 4.
Bimonthly
Order from: PLPC-OP
1.a; 2.a; 3.a; 4.a; 5.a; 6.a;
7. translation ceased publication.

EXPLANATION FOR PART I ENTRIES

** *Indicates that the title has been announced for publication in translation, but has not been so published as of the date of this study (June 1, 1972). Consult publisher for further information.*

Trans. begins — refers to year and volume of the original publication. Unless otherwise indicated, translated volume commences with issue number 1.

Trans. covers — refers to volume (or volumes) translated before translation ceased publication.

OP — following publisher code indicates back issues no longer available from the publisher.

<u>*Explanation of Coded data:*</u>

1. *a — Cover-to-cover translation*
   b — Selected articles translated

2. *a — Full translations of all articles*
   b — Some articles fully translated, others abridged or abstracted
   c — Selected articles abridged or abstracted
   d — Articles translated are fully translated, others merely cited or listed

3. *a — Translation carries original language numbers*
   b — Translation carries other numbers
   c — Translation carries no numbers

4. *a — Translation carries original language dates*
   b — Translation carries other dates
   c — Translation carries no dates

5. *a — Translation identifies original pagination*
   b — Translation does not identify original pagination

6. *a — Translation has volume index or collective table of contents*
   b — Translation has no volume index or collective table of contents

7. *"Lag time" between month of issue of original publication and month of issue of translated counter-part (there is a possible variation of one month because of mailing practices both of the original and the translation)*

8. *Other information (specified)*

9. *Translation exists but has not been located for verification and evaluation*

13 **APTECHNOE DELO**
(Pharmaceutics)
Trans. covers 1966, v. 15, nos. 1—5.
Bimonthly
Order from: NTIS
9.

14 **ARCHIVUM IMMUNOLOGIAE ET THERAPIAE EXPERIMENTALIS**
(Archives of Immunology and Experimental Therapy)
Trans. begins with 1962, v. 10+
Quarterly 1962—1963,
Bimonthly 1964+
Order from: NTIS
1.a; 2.a; 3.a; 4.a; 5.b; 6. ;
7. 5 mos.

15 **ARHIV BIOLOSKIKH NAUKA**
(Archives of Biological Sciences)
Trans. begins with 1962, v. 14+
Semiannual
Order from: NTIS
1.a; 2.a; 3.a; 4.a; 5.a; 6.b;
7. 30 mos.

16 **ARHIV ZA POLJOPRIVREDNE NAUKE I TEHNIKU**
(Journal of Scientific Agricultural Research)
Trans. covers 1962, v. 15.
4—5 issues per yr.
Order from: NTIS
9.

17 **ASTROFIZIKA**
(Astrophysics)
Trans. begins with 1965, v. 1+
Quarterly
Order from: PLPC
1.a; 2.a; 3.a; 4.a; 5.a; 6.a;
7. the present publisher of this translation is making every effort to reduce previously existing translation delays and it is deemed advisable not to cite any specific "lag time" in this edition of the GUIDE.

18 **ASTRONOMICHESKII VESTNIK**
(Solar Systems Research)
Trans. begins with 1967, v. 1+
Quarterly
Order from: PLPC
1.a; 2.a; 3.a; 4.a; 5.a; 6.a;
7. 9 mos.

19 **ASTRONOMICHESKII ZHURNAL**
(Soviet Astronomy—AJ)
Trans. begins with 1957, v. 34+
Bimonthly
Order from: AIP
1.a; 2.a; 3.a; 4.a; 5.a; 6.a;
7. 6 mos.

20 **ATOMNAYA ENERGIYA**
(Soviet Atomic Energy)
Trans. begins with 1956, v. 1+
Monthly
Order from: PLPC
1.a; 2.a; 3.a; 4.a; 5.a; 6.a;
7. 6 mos.;
8. vols. 1—13 were translated as "Soviet Journal of Atomic Energy" and a supplement was also issued for v. 2, 1957; selected articles were also published in the "Journal of Nuclear Energy" for the period 1956—1966; "Reactor Science and Technology" also lists the contents of coming issues and includes abstracts of the major articles; a French translation was also published, "Energie Atomique", covering 1963—1970, vols. 14—29, no. 6, by CNRS.

21 **AVIATSIYA I KOSMONAVTIKA**
(Aviation and Cosmonautics)
Order from: FTD
1.a; 2.a; 3.a; 4.a; 5.b; 6. ;
7. irregular;
8. restricted distribution, consult FTD for details.

22 **AVTOMATICHESKAYA SVARKA**
(Automatic Welding)
Trans. begins with 1959, v. 12+
Monthly
Order from: BWRA
1.a; 2.a; 3.a; 4.a; 5.a; 6.a;
7. 6 mos.

23 **AVTOMATIKA**
(Soviet Automatic Control)
Trans. begins with 1968+
Bimonthly
Order from: IEEE or ST
8. PP lists trans. and publication of v. 4, no. 3 in 1967;
9. IEEE issue not examined.

EXPLANATION FOR PART I ENTRIES

** *Indicates that the title has been announced for publication in translation, but has not been so published as of the date of this study (June 1, 1972). Consult publisher for further information.*

Trans. begins — refers to year and volume of the original publication. Unless otherwise indicated, translated volume commences with issue number 1.

Trans. covers — refers to volume (or volumes) translated before translation ceased publication.

OP — following publisher code indicates back issues no longer available from the publisher.

Explanation of Coded data:

1. a — Cover-to-cover translation
   b — Selected articles translated

2. a — Full translations of all articles
   b — Some articles fully translated, others abridged or abstracted
   c — Selected articles abridged or abstracted
   d — Articles translated are fully translated, others merely cited or listed

3. a — Translation carries original language numbers
   b — Translation carries other numbers
   c — Translation carries no numbers

4. a — Translation carries original language dates
   b — Translation carries other dates
   c — Translation carries no dates

5. a — Translation identifies original pagination
   b — Translation does not identify original pagination

6. a — Translation has volume index or collective table of contents
   b — Translation has no volume index or collective table of contents

7. "Lag time" between month of issue of original publication and month of issue of translated counter-part (there is a possible variation of one month because of mailing practices both of the original and the translation)

8. Other information (specified)

9. Translation exists but has not been located for verification and evaluation

24 AVTOMATIKA I TELEMEKHANIKA
(Automation and Remote Control)
Trans. begins with 1956, v. 17+
Monthly
Order from: PLPC
1.a; 2.a; 3.a; 4.a; 5.a; 6.a;
7. 6 mos.

25 AVTOMATIKA I VYCHISLITEL'NAYA TEKHNIKA
(Automatic Control)
Trans. begins with 1969, v. 3+
Bimonthly
Order from: APr
1.a; 2.a; 3.a; 4.a; 5.a; 6.a;
7. the present publisher of this translation is making every effort to reduce previously existing translation delays and it is deemed advisable not to cite any specific "lag time" in this edition of the GUIDE.

26 AVTOMETRIYA
(Autometry)
Trans. covers 1966–1970.
Bimonthly
Order from: SIC
1.a; 2.a; 3.a; 4.a; 5.a; 6.a;
8. SIC has not announced schedule for 1971+.

27 BILTEN INSTITUTA ZA NUKLEARNE NAUKE "BORIS KIDRIC"
(Bulletin of the Boris Kidric Institute of Nuclear Sciences)
Trans. covers 1962–1966, v. 13–17, no. 4.
Quarterly
Order from: NTIS
1.a; 2.a; 3.a; 4.a; 5.a; 6. ;
7. issued irregularly;
8. published as an AEC translation issued by NTIS.

28 BIOFIZIKA
(Biophysics)
Trans. begins with 1957, v. 2+
Bimonthly
Order from: PP
1.a; 2.a; 3.a; 4.a; 5.a; 6.a;
7. 12 mos.

29 BIOKHIMIYA
(Biochemistry)
Trans. begins with 1956, v. 21+
Bimonthly
Order from: PLPC
1.a; 2.a; 3.a; 4.a; 5.a; 6.a;
7. 6 mos.

30 BIOLOSKI GLASNIK
(Biological Review)
Trans. covers 1962, v. 15.
Quarterly
Order from: NTIS
7. translation ceased publication;
9.

31 BYULLETEN' EKSPERIMENTAL'NOI BIOLOGII I MEDITSINY
(Bulletin of Experimental Biology and Medicine)
Trans. begins with 1956, v. 41+
Monthly
Order from: PLPC
1.a; 2.a; 3.a; 4.a; 5.a; 6.a;
7. 6 mos.

BYULLETEN' IZOBRETENII see "Byulleten' Izobretenii i Tovarnykh Znakov"

32 BYULLETEN' IZOBRETENII I TOVARNYKH ZNAKOV
(Soviet Inventions Illustrated)
Issued in 3 sections; 1–Chemical; 2–Electrical; 3–Mechanical and General.
Trans. begins with 1959, v. 36+
Monthly
Order from: DIS
1.a; 2.c; 3.c; 4.a; 5.b; 6.a;
7. 2 mos.;
8. all illustrations reproduced; originally published under the title "Byulleten' Izobretenii" and translated from v. 36, 1959: vols. 36–37, 1959–1960, published as "Russian Patent Bulletin–Chemistry and Fuels" by RIS, published as "Derwent Russian Patent Report–Chemistry and Allied Subjects" by DIS; vols. 36–37, also published by DIS as "Russian Patent Abstracts" (sometimes titled "Russian Patent Gazette"), v. 38–39, no. 2, 1961–1962, published as "USSR Bulletin of Patents and Inventions" by DIS.

EXPLANATION FOR PART I ENTRIES

** *Indicates that the title has been announced for publication in translation, but has not been so published as of the date of this study (June 1, 1972). Consult publisher for further information.*

Trans. begins — refers to year and volume of the original publication. Unless otherwise indicated, translated volume commences with issue number 1.

Trans. covers — refers to volume (or volumes) translated before translation ceased publication.

OP — following publisher code indicates back issues no longer available from the publisher.

<u>*Explanation of Coded data:*</u>

1. a — Cover-to-cover translation
   b — Selected articles translated

2. a — Full translations of all articles
   b — Some articles fully translated, others abridged or abstracted
   c — Selected articles abridged or abstracted
   d — Articles translated are fully translated, others merely cited or listed

3. a — Translation carries original language numbers
   b — Translation carries other numbers
   c — Translation carries no numbers

4. a — Translation carries original language dates
   b — Translation carries other dates
   c — Translation carries no dates

5. a — Translation identifies original pagination
   b — Translation does not identify original pagination

6. a — Translation has volume index or collective table of contents
   b — Translation has no volume index or collective table of contents

7. "Lag time" between month of issue of original publication and month of issue of translated counter-part (there is a possible variation of one month because of mailing practices both of the original and the translation)

8. Other information (specified)

9. Translation exists but has not been located for verification and evaluation

33 BYULLETEN' MOSKOVSKOGO OBSHCHESTVA ISPYTATELEI PRIRODY, OTDEL. GEOLOGICHESKII
(Bulletin of the Moscow Society for Natural Research, Geological Section)
Trans. covers 1956–1958, vols. 31–33.
Semiannual
Order from: NTIS
9.

34 DEFEKTOSKOPIYA
(The Soviet Journal of Non-Destructive Testing)
Trans. begins with 1965, v. 1+
Bimonthly
Order from: PLPC
1.a; 2.a; 3.a; 4.a; 5.a; 6.a;
7. 6 mos.
8. earlier issues carried the word "Defectoscopy" before the remainder of the translation title.

35 DENKI GAKKAI ZASSHI
(Electrical Engineering in Japan)
Trans. begins with 1963, v. 83+
Monthly
Order from: IEEE or ST
1.b; 2.a; 3.a; 4.a; 5.b; 6.a;
7. 12 mos.;
8. research articles are translated, notes and abstracts are not.

36 DENKI KAGAKU (KYOKAI)
(Journal of the Electrochemical Society of Japan)
Trans. begins with 1960, v. 28+
Quarterly
Order from: ECJ
1.a; 2.b; 3.a; 4.a; 5.a; 6.a;
7. 6 mos.;
8. 1960–1962, vols. 28–30, are an almost complete translation of the original, v. 31+ translates selected articles in full and provides abstracts for all original articles (reviews of other papers and short 'technical communications' are omitted).

37 DENKI TSUSHIN GAKKAI ZASSHI
(Electronics and Communications in Japan)
Trans. covers 1963–1967, vols. 46–50.
Monthly
Order from: IEEE
1.a; 2.a; 3.a; 4.a; 5.b; 6.b;
8. see DENSHI TSUSHIN GAKKAI RONBUNSHI for successor to translated title.

38 DENSHI TSUSHIN GAKKAI RONBUNSHI
(Electronics and Communications in Japan)
Trans. begins with 1968, v. 51+
Monthly
Order from: IEEE or ST
1.a; 2.a; 3.a; 4.a; 5.b; 6.b;
7. 12 mos.
8. same translation title was previously used for translations of DENKI TSUSHIN GAKKAI ZASSHI.

39 DEREVOOBRABATYVAYUSHCHAYA PROMYSHLENNOST'
(Industrial Wood Processing)
Trans. covers 1959–1960, v. 8, no. 7 – v. 9.
Monthly
Order from: NLL
7. translation ceased publication;
9.

40 DIFFERENTSIAL'NYE URAVNENIYA
(Differential Equations)
Trans. begins with 1965, v. 1+
Monthly
Order from: PLPC
1.a; 2.a; 3.a; 4.a; 5.a; 6.a;
7. the present publisher of this translation is making every effort to reduce previously existing translation delays and it is deemed advisable not to cite any specific "lag time" in this edition of the GUIDE.

DOKLADY AKADEMII NAUK SSSR
[Note: while the Soviet publication is a single journal, various sections are translated by different publishers under different titles. When doubt exists concerning the articles desired, a call to a library possessing the original Soviet publication will quickly identify the subject classification comparable to that used by the appropriate translation journal in its title.]

41 (Doklady Biochemistry)
Trans. begins with 1957, v. 112+
Bimonthly
Order from: PLPC
1.a; 2.a; 3.a; 4.a; 5.a; 6.a;
7. 6 mos.
8. vols. 118–141, 1958–1961, are available from NTIS, vols. 142–153, 1962–1963, were included in "Doklady Biological Sciences" in translation and are available from PLPC.

EXPLANATION FOR PART I ENTRIES

** *Indicates that the title has been announced for publication in translation, but has not been so published as of the date of this study (June 1, 1972). Consult publisher for further information.*

Trans. begins — refers to year and volume of the original publication. Unless otherwise indicated, translated volume commences with issue number 1.

Trans. covers — refers to volume (or volumes) translated before translation ceased publication.

OP — following publisher code indicates back issues no longer available from the publisher.

<u>*Explanation of Coded data:*</u>

1. *a — Cover-to-cover translation*
   b — Selected articles translated

2. *a — Full translations of all articles*
   b — Some articles fully translated, others abridged or abstracted
   c — Selected articles abridged or abstracted
   d — Articles translated are fully translated, others merely cited or listed

3. *a — Translation carries original language numbers*
   b — Translation carries other numbers
   c — Translation carries no numbers

4. *a — Translation carries original language dates*
   b — Translation carries other dates
   c — Translation carries no dates

5. *a — Translation identifies original pagination*
   b — Translation does not identify original pagination

6. *a — Translation has volume index or collective table of contents*
   b — Translation has no volume index or collective table of contents

7. *"Lag time" between month of issue of original publication and month of issue of translated counter-part (there is a possible variation of one month because of mailing practices both of the original and the translation)*

8. *Other information (specified)*

9. *Translation exists but has not been located for verification and evaluation*

42 (Doklady Biological Sciences)
Trans. begins with 1957, v. 112+
Bimonthly
Order from: PLPC
1.a; 2.a; 3.a; 4.a; 5.a; 6.a;
7. 6 mos.
8. vols. 112–141, 1957–1961, were translated by and are available from NTIS; beginning with 1964, v. 154+ all sections are included except biochemistry, biophysics and botany.

43 (Doklady Biophysics)
Trans. begins with 1957, v. 112+
Bimonthly
Order from: PLPC
1.a; 2.a; 3.a; 4.a; 5.a; 6.a;
7. 6 mos.;
8. each translated issue covers 3 original volumes; vols. 112–141 (1957–1961) are included in "Doklady Biological Sciences", available from NTIS; vols. 142–153 (1962–1963) were included in "Doklady Biological Sciences" available from PLPC; v. 154+ (1964+) issued as a separate journal by PLPC.

44 (Doklady Botanical Sciences)
Trans. begins with 1957, v. 112+
Bimonthly
Order from: PLPC
1.a; 2.a; 3.a; 4.a; 5.a; 6.a;
7. 8 mos.
8. one translated issue covers two original volumes; vols. 112–141 (1957–1961) are included in "Doklady Biological Sciences", available from NTIS; vols. 142–153 (1962–1963) were included in "Doklady Biological Sciences" available from PLPC; v. 154+ (1964+) issued as a separate journal by PLPC.

45 (Doklady Chemical Technology)
Trans. begins with 1956, v. 106+
3 issues per year
Order from: PLPC
1.a; 2.a; 3.a; 4.a; 5.a; 6.a;
7. 6 mos.
8. one translated volume covers two original volumes.

46 (Doklady Chemistry)
Trans. begins with 1956, v. 106+
Bimonthly
Order from: PLPC
1.a; 2.a; 3.a; 4.a; 5.a; 6.a;
7. 6 mos.

[Oceanology] for Doklady sections see "Soviet Oceanography"

47 (Doklady of the Academy of Sciences of the USSR, Earth Science Sections)
Trans. begins with 1959, v. 124+
Bimonthly
Order from: AGI
1.a; 2.a; 3.a; 4.a; 5.a; 6.b;
7. 8 mos.;
8. vols. 124–141 (1959–1961) are available from KP; see "Proc. Acad. Sci. USSR, Geochemistry Section" for earlier translated volumes of the geochemistry sections; see "Proc. Acad. Sci. USSR, Geology" for earlier volumes of the geology sections.

48 (Doklady Physical Chemistry)
Trans. begins with 1957, v. 112+
Bimonthly
Order from: PLPC
1.a; 2.a; 3.a; 4.a; 5.a; 6.a;
7. 6 mos.

49 (Proceedings of the Academy of Sciences of the USSR. Agrochemistry Section)
Trans. covers 1956, vols. 106–111.
Bimonthly
Order from: PLPC-OP
7. translation ceased publication.
9.

50 (Proceedings of the Academy of Sciences of the USSR. Applied Physics Section)
Trans. covers 1957–1958, vols. 112–123.
Bimonthly
Order from: PLPC-OP
1.a; 2.a; 3.a; 4.a; 5.a; 6.a;
7. translation ceased publication.

EXPLANATION FOR PART I ENTRIES

**   *Indicates that the title has been announced for publication in translation, but has not been so published as of the date of this study (June 1, 1972). Consult publisher for further information.*

Trans. begins — refers to year and volume of the original publication. Unless otherwise indicated, translated volume commences with issue number 1.

Trans. covers — refers to volume (or volumes) translated before translation ceased publication.

OP — following publisher code indicates back issues no longer available from the publisher.

Explanation of Coded data:

1. a — *Cover-to-cover translation*
   b — *Selected articles translated*

2. a — *Full translations of all articles*
   b — *Some articles fully translated, others abridged or abstracted*
   c — *Selected articles abridged or abstracted*
   d — *Articles translated are fully translated, others merely cited or listed*

3. a — *Translation carries original language numbers*
   b — *Translation carries other numbers*
   c — *Translation carries no numbers*

4. a — *Translation carries original language dates*
   b — *Translation carries other dates*
   c — *Translation carries no dates*

5. a — *Translation identifies original pagination*
   b — *Translation does not identify original pagination*

6. a — *Translation has volume index or collective table of contents*
   b — *Translation has no volume index or collective table of contents*

7. *"Lag time" between month of issue of original publication and month of issue of translated counter-part (there is a possible variation of one month because of mailing practices both of the original and the translation)*

8. *Other information (specified)*

9. *Translation exists but has not been located for verification and evaluation*

51 (Proceedings of the Academy of Sciences of the USSR. Geochemistry Section)
Trans. covers 1956–1958, vols. 106–123.
Order from: PLPC
1.a; 2.a; 3.a; 4.a; 5.a; 6.a;
7. translation ceased separate publication;
8. see "Doklady of the Academy of Sciences of the USSR. Earth Science Section" for later issues.

52 (Proceedings of the Academy of Sciences of the USSR. Geological Sciences Section)
Trans. covers 1957–1958, vols. 112–123.
Bimonthly
Order from: PLPC
1.a; 2.a; 3.a; 4.a; 5.a; 6.a;
7. translation ceased separate publication;
8. see "Doklady of the Academy of Sciences of the USSR. Earth Science Section" for later issues.

53 (Proceedings of the Academy of Sciences of the USSR. Soil Science Section)
Trans. of 1961, v. 131.
Order from: NTIS
Trans. of 1962, v. 132.
Order from: ST
9.

54 (Soviet Mathematics–Doklady)
Trans. begins with 1960, v. 130+
Bimonthly
Order from: AMS
1.a; 2.a; 3.b; 4.a; 5.a; 6.a;
7. 5 mos.

55 (Soviet Physics–Doklady)
Trans. begins with 1956, v. 106+
Monthly
Order from: AIP
1.a; 2.a; 3.b; 4.b; 5.a; 6.a;
7. 6 mos.
8. each translated issue covers one volume of the original physics section.

56 DRAHT FACHZEITSCHRIFT
(Wire: Coburg.)
Trans. begins with 1951+
Monthly
Order from: DFC
1.b; 2.a; 3.a; 4.a; 5.b; 6.a;
7. 1 month;
[continued next column]

8. other editions: "Le Trefile" (French, bimonthly); "Il Filo Metallico" (Italian, quarterly); "Alambre" (Spanish, quarterly); all non-English editions are 9.

57 EKOLOGIYA
(Soviet Journal of Ecology)
Trans. begins with 1970, v. 1+
Bimonthly
Order from: PLPC
1.a; 2.a; 3.a; 4.a; 5.a; 6.a;
7. 10 mos.
8. in the years 1970–1971, the translation was simply titled "Ecology".

58 ELEKTRICHESKIE STANTSII
(Soviet Power Engineering)
Trans. begins with 1971, no. 1+
Monthly
Order from: McE
1.a; 2.a; 3.a; 4.a; 5.a; 6. ;
7.

59 ELEKTRICHESTVO
(Electric Technology USSR)
Trans. begins with 1957, v. 77+
Quarterly
Order from: PP
1.a; 2.b; 3.b; 4.a; 5.a; 6.b;
7. 12 mos.;
8. selected articles are translated in full, others are abstracted.

60 ELEKTROKHIMIYA
(Soviet Electrochemistry)
Trans. begins with 1965, v. 1+
Monthly
Order from: PLPC
1.a; 2.a; 3.a; 4.a; 5.a; 6.a;
7. 6 mos.

61 ELEKTRONNAYA OBRABOTKA MATERIALOV
(Applied Electrical Phenomena)
Trans. covers 1965–1969.
Bimonthly
Order from: SIC
1.a; 2.a; 3.a; 4.a; 5.a; 6.a;
8. SIC has not announced schedule for 1970+.

EXPLANATION FOR PART I ENTRIES

** *Indicates that the title has been announced for publication in translation, but has not been so published as of the date of this study (June 1, 1972). Consult publisher for further information.*

Trans. begins — refers to year and volume of the original publication. Unless otherwise indicated, translated volume commences with issue number 1.

Trans. covers — refers to volume (or volumes) translated before translation ceased publication.

OP — following publisher code indicates back issues no longer available from the publisher.

<u>*Explanation of Coded data:*</u>

1. *a — Cover-to-cover translation*
   b — Selected articles translated

2. *a — Full translations of all articles*
   b — Some articles fully translated, others abridged or abstracted
   c — Selected articles abridged or abstracted
   d — Articles translated are fully translated, others merely cited or listed

3. *a — Translation carries original language numbers*
   b — Translation carries other numbers
   c — Translation carries no numbers

4. *a — Translation carries original language dates*
   b — Translation carries other dates
   c — Translation carries no dates

5. *a — Translation identifies original pagination*
   b — Translation does not identify original pagination

6. *a — Translation has volume index or collective table of contents*
   b — Translation has no volume index or collective table of contents

7. *"Lag time" between month of issue of original publication and month of issue of translated counter-part (there is a possible variation of one month because of mailing practices both of the original and the translation)*

8. *Other information (specified)*

9. *Translation exists but has not been located for verification and evaluation*

62 ELEKTROSVYAZ'
(Telecommunications)
Trans. covers 1957–1962, vols. 12–16.
Monthly
Order from: IEEE or ST
1.a; 2.a; 3.a; 4.a; 5.b; 6.a;
7. 7 mos.;
8. translation ceased separate publication, became incorporated with "Radiotekhnika" (in translation) as "Telecommunications and Radio Engineering" beginning v. 17, 1963.

63 ELEKTROTEKHNIKA
(Soviet Electrical Engineering)
Trans. begins with 1965, v. 36+
Monthly
Order from: FP
1.a; 2.a; 3.a; 4.a; 5.a; 6.a;
8. translation ceased publication, specific termination date unknown.

64 ENDOKRYNOLOGIA POLSKA
(Polish Endrocrinology)
Trans. begins with 1962, v. 13+
Annual 1962, 3 issues per annum 1963+, 6 issues per annum 1970+
Order from: NTIS
1.b; 2.a; 3.a; 4.a; 5.b; 6.a;
7. 12 mos.

65 ENTOMOLOGICHESKOE OBOZRENIE
(Entomological Review)
Trans. begins with 1958, v. 37+
Quarterly
Order from: ST
1.a; 2.a; 3.a; 4.a; 5.a; 6.a;
7. 9 mos.;
8. vols. 37–40 are available from NTIS.

66 FARMAKOLOGIYA I TOKSIKOLOGIYA
(Russian Pharmacology and Toxicology)
Trans. begins with 1967, v. 30+
Bimonthly
Order from: Euromed
1.a; 2.a; 3.a; 4.a; 5.a; 6.a;
7.
8. vols. 20–22, 1957–1959 were translated as "Pharmacology and Toxicology" by PLPC-OP.

67 FIZIKA ELEMENTARNYKH CHASTITS I ATOMNOGO YADRA**
(Soviet Journal of Particles and Nuclei)
Trans. to begin with 1972, v. 3+
Quarterly
Order from: AIP
8. vols. 1 and 2, published as books in the USSR, are in the process of being translated as annuals by PLPC.

FIZIKA ZEMLI see
"Izvestiya Akad. Nauk SSSR. Fizika Zemli"

FIZIKI ATMOSFERY I OKEANA see
"Izvestiya Akad. Nauk SSSR. Fiziki Atmosfery I Okeana"

68 FIZIKA GORENIYA I VZRYVA
(Combustions, Explosion, and Shock Waves)
Trans. begins with 1965, v. 1+
Quarterly
Order from: PLPC
1.a; 2.a; 3.a; 4.a; 5.a; 6.a;
7. the present publisher of this translation is making every effort to reduce previously existing translation delays and it is deemed advisable not to cite any specific "lag time" in this edition of the GUIDE.
8. numbers for 1965–1966 of the original had the title "Nauchnotekhnicheskii problemy goreniya i vzryva".

69 FIZIKA I TEKHNIKA POLUPROVODNIKOV
(Soviet Physics—Semiconductors)
Trans. begins with 1967, v. 1+
Monthly
Order from: AIP
1.a; 2.a; 3.a; 4.a; 5.a; 6.a;
7. 6 mos.

70 FIZIKA METALLOV I METALLOVEDENIE
(Physics of Metals and Metallography)
Trans. begins with 1957, v. 4+
Monthly
Order from: PP
1.a; 2.a; 3.a; 4.a; 5.a; 6.a;
7. 15 mos.

EXPLANATION FOR PART I ENTRIES

** *Indicates that the title has been announced for publication in translation, but has not been so published as of the date of this study (June 1, 1972). Consult publisher for further information.*

Trans. begins — refers to year and volume of the original publication. Unless otherwise indicated, translated volume commences with issue number 1.

Trans. covers — refers to volume (or volumes) translated before translation ceased publication.

OP — following publisher code indicates back issues no longer available from the publisher.

Explanation of Coded data:

1. a — Cover-to-cover translation
   b — Selected articles translated

2. a — Full translations of all articles
   b — Some articles fully translated, others abridged or abstracted
   c — Selected articles abridged or abstracted
   d — Articles translated are fully translated, others merely cited or listed

3. a — Translation carries original language numbers
   b — Translation carries other numbers
   c — Translation carries no numbers

4. a — Translation carries original language dates
   b — Translation carries other dates
   c — Translation carries no dates

5. a — Translation identifies original pagination
   b — Translation does not identify original pagination

6. a — Translation has volume index or collective table of contents
   b — Translation has no volume index or collective table of contents

7. "Lag time" between month of issue of original publication and month of issue of translated counter-part (there is a possible variation of one month because of mailing practices both of the original and the translation)

8. Other information (specified)

9. Translation exists but has not been located for verification and evaluation

71 FIZIKA TVERDOGO TELA
(Soviet Physics–Solid State)
> Trans. begins with 1959, v. 1+
> Monthly
> Order from: AIP
> 1.a; 2.a; 3.a; 4.a; 5.a; 6.a;
> 7. 6 mos.

72 FIZIKO-KHIMICHESKAYA MEKHANIKA MATERIALOV
(Soviet Materials Science)
> Trans. begins with 1965, v. 1+
> Bimonthly
> Order from: PLPC
> 1.a; 2.a; 3.a; 4.a; 5.a; 6.a;
> 7. the present publisher of this translation is making every effort to reduce previously existing translation delays and it is deemed advisable not to cite any specific "lag time" in this edition of the GUIDE.

73 FIZIKO-TEKHNICHESKIE PROBLEMY RAZRABOTKI POLEZNYKH ISKOPAEMYKH
(Soviet Mining Science)
> Trans. begins with 1965, v. 1+
> Bimonthly
> Order from: PLPC
> 1.a; 2.a; 3.a; 4.a; 5.a; 6.a;
> 7. 6 mos.

74 FIZIOLOGICHESKII ZHURNAL SSSR IMENI I. M. SECHENOVA
(Sechenov Physiological Journal of the USSR)
> Trans. covers 1957–1961, vols. 43–47.
> Order from: PP
> 7. translation ceased publication.
> 8. v. 47, 1961 also translated by ST under same title.
> 9.

75 FIZIOLOGIYA I BIOKHIMIYA KUL'TURNYK RASTENII
(Physiology and Biochemistry of Cultivated Plants)
> Trans. covers 1969–1970, vols. 1–2.
> Bimonthly
> Order from: PLPC
> 1.a; 2.a; 3.a; 4.a; 5.a; 6.a;
> 7. translation ceased publication.

76 FIZIOLOGIYA RASTENII
(Soviet Plant Physiology)
> Trans. begins with 1957, v. 4+
> Bimonthly
> Order from: PLPC
> 1.a; 2.a; 3.a; 4.a; 5.a; 6.a;
> 7. 6 mos.
> 8. vols. 4–8, 1957–1961, published by NTIS.

77 FOLIA MORPHOLOGICA
> Trans. begins with 1963, v. 14+
> Quarterly
> Order from: NTIS
> 1.a; 2.a; 3.a; 4.a; 5.b; 6.a;
> 7. 9 mos.;
> 8. another publication, "Ceskoslovenskaya Morphologica" (later changed to "Folia Morphologia") is published in English by the Czech. Acad. of Sci.

78 FUNKTSIONAL'NYI ANALYZ I EGO PRILOZHENIYA
(Functional Analysis and Its Applications)
> Trans. begins with 1967, no. 1+
> Quarterly
> Order from: PLPC
> 1.a; 2.a; 3.a; 4.a; 5.a; 6. ;
> 7. 9 mos.

79 GELIOTEKHNIKA
(Applied Solar Energy)
> Trans. begins with 1965, v. 1+
> Bimonthly
> Order from: APr
> 1.a; 2.a; 3.a; 4.a; 5.a; 6.a;
> 7. the present publisher of this translation is making every effort to reduce previously existing translation delays and it is deemed advisable not to cite any specific "lag time" in this edition of the GUIDE.
> 8. another translation, "Heliotechnology", beginning with 1966, v. 2+, monthly, is available from NTIS.

80 GENETIKA
(Soviet Genetics)
> Trans. begins with 1966, v. 2+
> Monthly
> Order from: PLPC
> 1.a; 2.a; 3.a; 4.a; 5.a; 6.a;
> [continued next page]

EXPLANATION FOR PART I ENTRIES

** *Indicates that the title has been announced for publication in translation, but has not been so published as of the date of this study (June 1, 1972). Consult publisher for further information.*

Trans. begins — refers to year and volume of the original publication. Unless otherwise indicated, translated volume commences with issue number 1.

Trans. covers — refers to volume (or volumes) translated before translation ceased publication.

OP — following publisher code indicates back issues no longer available from the publisher.

Explanation of Coded data:

1. *a — Cover-to-cover translation*
   b — Selected articles translated

2. *a — Full translations of all articles*
   b — Some articles fully translated, others abridged or abstracted
   c — Selected articles abridged or abstracted
   d — Articles translated are fully translated, others merely cited or listed

3. *a — Translation carries original language numbers*
   b — Translation carries other numbers
   c — Translation carries no numbers

4. *a — Translation carries original language dates*
   b — Translation carries other dates
   c — Translation carries no dates

5. *a — Translation identifies original pagination*
   b — Translation does not identify original pagination

6. *a — Translation has volume index or collective table of contents*
   b — Translation has no volume index or collective table of contents

7. *"Lag time" between month of issue of original publication and month of issue of translated counter-part (there is a possible variation of one month because of mailing practices both of the original and the translation)*

8. *Other information (specified)*

9. *Translation exists but has not been located for verification and evaluation*

7. the present publisher of this translation is making every effort to reduce previously existing translation delays and it is deemed advisable not to cite any specific "lag time" in this edition of the GUIDE.
8. another translation, "Genetics", announced by NTIS was discontinued, only 1965, no. 1 and 1966, no. 12 were published.

81 GEODEZIYA I KARTOGRAFIYA
(Geodesy and Cartography)
*Trans. covers 1959–1961, vols. 4–7.
Bimonthly
Order from: AGU
7. translation ceased publication.
9.*

82 GEOKHIMIYA
(Geochemistry)
*Trans. covers 1956–1963, vols. 1–8.
Monthly
Order from: AGI
1.a; 2.a; 3.b; 4.a; 5.a; 6.a;
7. translation ceased publication.
8. commencing 1964, no. 1+, selected articles are included in "Geochemistry International" (some articles are translated in full, some are abstracted), articles not translated in GI are available from AGI at nominal charge.*

83 GEOLOGIYA NEFTI I GAZA
(Petroleum Geology)
*Trans. begins with 1958, v. 2+
Monthly
Order from: PG
1.a; 2.a; 3.a; 4.a; 5.b; 6.b;
7. 30 mos.*

84 GEOLOGIYA RUDNYKH MESTOROZHDENII
(Economic Geology USSR)
*Trans. begins with 1960, no. 1+
Order from: PP
1.a; 2.b; 3.a; 4.a; 5.a; 6.a;
7. 24 mos.;
8. beginning with 1959, no. 1+, reviews and excerpts are included in the 'Reviews' section of "Economic Geology", available from EGP; see "Litologiya i Poleznye Iskopaemye" for a parallel source.*

85 GEOMAGNETIZM I AERONOMIYA
(Geomagnetism and Aeronomy)
*Trans. begins with 1961, v. 1+
Bimonthly
Order from: AGU
1.a; 2.a; 3.a; 4.a; 5.a; 6.a;
7. 8 mos.*

86 GEOTEKHTONIKA
(Geotectonics)
*Trans. begins with 1967, v. 2+
Bimonthly
Order from: AGU
1.a; 2.a; 3.a; 4.a; 5.a; 6.a;
7. 8 mos.*

87 GIDROBIOLOGICHESKII ZHURNAL
(Hydrobiological Journal)
*Trans. begins with 1969, v. 5+
Bimonthly
Order from: AFS
9.*

88 GIDROTEKHNICHESKOE STROITEL'STVO
(Hydrotechnical Construction)
*Trans. begins with 1967, no. 1+
Order from: ASCE
1.a; 2.a; 3.a; 4.a; 5.a; 6. ;
7. 6 mos.*

89 GIGIENA I SANITARIYA
(Hygiene and Sanitation)
*Trans. begins with 1964, v. 29+
Monthly 1964, quarterly 1965+
Order from: NTIS
1.a; 2.a; 3.a; 4.a; 5.b; 6.a;
7. 12 mos.*

90 GIGIENA TRUDA I PROFESSIONAL'NYE ZABOLEVANIYA
(Labor Hygiene and Occupational Diseases)
*Trans. covers 1966, v. 10.
Monthly
Order from: NTIS
9.*

91 GLASNIK SRPSKO KHEMIJSKOG DRUSTVA
(Bulletin of the Chemical Society, Belgrade)
*Trans. begins with 1962, v. 27+
10 issues per year
Order from: NTIS
9.*

EXPLANATION FOR PART I ENTRIES

** *Indicates that the title has been announced for publication in translation, but has not been so published as of the date of this study (June 1, 1972). Consult publisher for further information.*

Trans. begins — *refers to year and volume of the original publication. Unless otherwise indicated, translated volume commences with issue number 1.*

Trans. covers — *refers to volume (or volumes) translated before translation ceased publication.*

OP — *following publisher code indicates back issues no longer available from the publisher.*

Explanation of Coded data:

1. a — Cover-to-cover translation
   b — Selected articles translated

2. a — Full translations of all articles
   b — Some articles fully translated, others abridged or abstracted
   c — Selected articles abridged or abstracted
   d — Articles translated are fully translated, others merely cited or listed

3. a — Translation carries original language numbers
   b — Translation carries other numbers
   c — Translation carries no numbers

4. a — Translation carries original language dates
   b — Translation carries other dates
   c — Translation carries no dates

5. a — Translation identifies original pagination
   b — Translation does not identify original pagination

6. a — Translation has volume index or collective table of contents
   b — Translation has no volume index or collective table of contents

7. "Lag time" between month of issue of original publication and month of issue of translated counter-part (there is a possible variation of one month because of mailing practices both of the original and the translation)

8. Other information (specified)

9. Translation exists but has not been located for verification and evaluation

92 HUA HSUEH HSUEH PAO
(Acta Chimica Sinica)
Trans. covers 1966, v. 32, nos. 1–3
Order from: PLPC
1.a; 2.a; 3.a; 4.a; 5.a; 6.a;
7. original ceased publication 1967.

IADERNAIA FIZIKA see
"Yadernaya Fizika"

93 IGAKU CHUO ZASSHI
(Abstracts of Japanese Medicine)
Trans. covers 1960–1962, vols. 1–2.
Monthly
Order from: EMF
7. translation ceased publication.
9.

94 INFORMATSIONNYI BYULLETEN' SOVETSKOI ANTARTICHESKOI EKSPEDITSII
(Soviet Antarctic Expedition Information Bulletin)
Trans. covers 1961–1964, v. 4, no. 31– v. 7, no. 78.
Bimonthly
Order from: AGU
1.a; 2.a; 3.a; 4.a; 5.a; 6.a;
8. for 1958–1961, issues 1–30, see "Sovetskaya Antarticheskaya Ekspeditsiya, Informatsionnyi Byulleten' ".

95 INZHENERNO-FIZICHESKII ZHURNAL
(Journal of Engineering Physics)
Trans. begins with 1962, v. 5+
Monthly
Order from: PLPC
1.a; 2.a; 3.a; 4.a; 5.a; 6.a;
7. the present publisher of this translation is making every effort to reduce previously existing translation delays and it is deemed advisable not to cite any specific "lag time" in this edition of the GUIDE.
8. vols. 5–7, 1962–1964 (abstracts only) available from FP, vol. 6 (selected articles) available from NTIS.

96 INZHENERNYI ZHURNAL
(Soviet Engineering Journal)
Trans. covers 1963–1966, vols. 3–6.
Bimonthly
Order from: FP-OP
1.a; 2.a; 3.a; 4.a; 5.a; 6.a;
[continued next column]

8. v. 3 (1963), nos. 1–4, v. 4 (1964), nos. 1–2, 4, and v. 5 (1965), nos. 1 & 2, were translated under the title "Engineering Journal" and are available from NTIS; the FP translation covers vols. 5–6; see "Inzhenernyi Zhurnal Mekhanika Tverdogo Tela" for successor publication (also translated).

97 INZHENERNYI ZHURNAL, MEKHANIKA TVERDOGO TELA
(Mechanics of Solids)
Trans. begins with 1967, v. 1+
Bimonthly
Order from: APr
1.a; 2.a; 3.a; 4.a; 5.a; 6.a;
7. the present publisher of this translation is making every effort to reduce previously existing translation delays and it is deemed advisable not to cite any specific "lag time" in this edition of the GUIDE.
8. see "Inzhenernyi Zhurnal" for preceeding Soviet title; commencing with 1969, the Russian title changes to "Izv. Akad. Nauk SSSR, Mekhanika Tverdogo Tela".

98 ISKUSSTVENNYE SPUTNIKI ZEMLI
(Artificial Earth Satellites)
Trans. covers 1960–1961, vols. 1–6.
Order from: PLPC
1.a; 2.a; 3.a; 4.a; 5.a; 6.a;
7. original and translation ceased publication;
8. see "Kosmicheskie Issledovaniya" (Cosmic Research) for successor publication.

99 ISSLEDOVANIYA OBLAKOV, OSADKOV I GROZOVOGO ELEKTRICHESTVA
(Studies of Clouds, Precipitation and Thunderstorm Electricity)
Trans. covers 1961.
Order from: AmMetSoc

100 IZMERITEL'NAYA TEKHNIKA
(Measurement Techniques)
Trans. begins with 1958, v. 7+
Bimonthly 1958, monthly 1959+
Order from: PLPC
1.a; 2.a; 3.a; 4.a; 5.a; 6.a;
7. 6 mos.

101 IZVESTIYA AKAD. NAUK KAZAKH. SSR, SERIYA KHIMICHESKAYA**
Trans. planned to begin with 1973, no. 1+
Order from: SP

EXPLANATION FOR PART I ENTRIES

** *Indicates that the title has been announced for publication in translation, but has not been so published as of the date of this study (June 1, 1972). Consult publisher for further information.*

Trans. begins — refers to year and volume of the original publication. Unless otherwise indicated, translated volume commences with issue number 1.

Trans. covers — refers to volume (or volumes) translated before translation ceased publication.

OP — following publisher code indicates back issues no longer available from the publisher.

<u>Explanation of Coded data:</u>

1. *a — Cover-to-cover translation*
   b — Selected articles translated

2. *a — Full translations of all articles*
   b — Some articles fully translated, others abridged or abstracted
   c — Selected articles abridged or abstracted
   d — Articles translated are fully translated, others merely cited or listed

3. *a — Translation carries original language numbers*
   b — Translation carries other numbers
   c — Translation carries no numbers

4. *a — Translation carries original language dates*
   b — Translation carries other dates
   c — Translation carries no dates

5. *a — Translation identifies original pagination*
   b — Translation does not identify original pagination

6. *a — Translation has volume index or collective table of contents*
   b — Translation has no volume index or collective table of contents

7. *"Lag time" between month of issue of original publication and month of issue of translated counter-part (there is a possible variation of one month because of mailing practices both of the original and the translation)*

8. *Other information (specified)*

9. *Translation exists but has not been located for verification and evaluation*

IZVESTIYA AKAD. NAUK SSSR. MEKHANIKA TVERDOGO TELA
(only the predecessor, "Inzhenernyi Zhurnal, Mekhanika Tverdogo Tela" is presently being translated)

102 **IZVESTIYA AKADEMII NAUK SSSR. MEKHANIKA ZHIDKOSTEI I GAZOV**
(Fluid Dynamics)
Trans. begins with 1966, v. 1+
Bimonthly
Order from: PLPC
1.a; 2.a; 3.a; 4.a; 5.a; 6.a;
7. the present publisher of this translation is making every effort to reduce previously existing translation delays and it is deemed advisable not to cite any specific "lag time" in this edition of the GUIDE.
8. original publication formerly titled "Izvestiya Akademii Nauk SSSR. Mekhanika" (translation previously announced as "Soviet Fluid Mechanics").

IZVESTIYA AKADEMII NAUK SSSR. METALLURGIYA I GORNOE DELO see "Izvestiya Akademii Nauk SSSR. Metally"

103 **IZVESTIYA AKADEMII NAUK SSSR. METALLY**
(Russian Metallurgy)
Trans. begins with 1960, no. 1+
Bimonthly
Order from: SIC
1.b; 2.b; 3.a; 4.a; 5.a; 6.a;
7. 12 mos.;
8. for 1963–1964 the Soviet title was "Izvestiya Akademii Nauk SSSR. Metallurgiya i Gornoe Delo", (translated as "Russian Metallurgy and Mining", available from SIC); before 1963 the Soviet title was "Izvestiya Akademii Nauk SSSR. Otdelenie Tekhnicheskikh Nauk. Metallurgiya i Toplivo" (translated as "Russian Metallurgy and Fuels", and available from ETP for 1960 and 1962, with translation of articles from the 1961 Soviet publication available from SIC).

IZVESTIYA AKADEMII NAUK SSSR. OTDELENIE KHIMICHESKIKH NAUK see "Izvestiya Akademii Nauk SSSR. Seriya Khimicheskaya"

IZVESTIYA AKADEMII NAUK SSSR. OTDELENIE TEKHNICHESKIKH NAUK. METALLURGIYA I TOPLIVO see "Izvestiya Akademii Nauk SSSR. Metally"

IZVESTIYA AKADEMII NAUK SSSR. OTDELENIE TEKHNICHESKIKH NAUK. TEKHNICHESKAYA KIBERNETIKA see "Izvestiya Akademii Nauk SSSR. Tekhnicheskaya Kibernetika"

104 **IZVESTIYA AKADEMII NAUK SSSR. SERIYA FIZICHESKAYA**
(Bulletin of the Academy of Sciences of the USSR. Physical Series)
Trans. begins with 1954, v. 18, no. 3+
Monthly
Order from: CTT
1.a; 2.a; 3.a; 4.a; 5.a; 6.a;
7. 12 mos.

105 **IZVESTIYA AKADEMII NAUK SSSR. FIZIKI ATMOSFERY I OKEANA**
(Atmospheric and Ocean Physics)
Trans. begins with 1965, no. 1+
Monthly
Order from: AGU
1.a; 2.a; 3.a; 4.a; 5.a; 6.a;
7. 7 mos.;
8. see "Izvestiya Akademii Nauk SSSR. Seriya Geofizicheskaya" (also translated) for preceeding Soviet publication.

106 **IZVESTIYA AKADEMII NAUK SSSR. FIZIKI ZEMLI**
(Physics of the Solid Earth)
Trans. begins with 1965, no. 1+
Monthly
Order from: AGU
1.a; 2.a; 3.a; 4.a; 5.a; 6.a;
7. 7 mos.;
8. preceeding Soviet publication was "Izvestiya Akademii Nauk SSSR. Seriya Geofizicheskaya" (also translated).

107 **IZVESTIYA AKADEMII NAUK SSSR. GEOFIZICHESKAYA**
(Bulletin. Academy of Sciences USSR. Geophysics Series)
Trans. covers 1957–1964.
Monthly
Order from: AGU
1.a; 2.a; 3.a; 4.a; 5.a; 6. ;
[continued next page]

EXPLANATION FOR PART I ENTRIES

**   *Indicates that the title has been announced for publication in translation, but has not been so published as of the date of this study (June 1, 1972). Consult publisher for further information.*

Trans. begins — refers to year and volume of the original publication. Unless otherwise indicated, translated volume commences with issue number 1.

Trans. covers — refers to volume (or volumes) translated before translation ceased publication.

OP — following publisher code indicates back issues no longer available from the publisher.

Explanation of Coded data:

1. a — Cover-to-cover translation
   b — Selected articles translated

2. a — Full translations of all articles
   b — Some articles fully translated, others abridged or abstracted
   c — Selected articles abridged or abstracted
   d — Articles translated are fully translated, others merely cited or listed

3. a — Translation carries original language numbers
   b — Translation carries other numbers
   c — Translation carries no numbers

4. a — Translation carries original language dates
   b — Translation carries other dates
   c — Translation carries no dates

5. a — Translation identifies original pagination
   b — Translation does not identify original pagination

6. a — Translation has volume index or collective table of contents
   b — Translation has no volume index or collective table of contents

7. "Lag time" between month of issue of original publication and month of issue of translated counter-part (there is a possible variation of one month because of mailing practices both of the original and the translation)

8. Other information (specified)

9. Translation exists but has not been located for verification and evaluation

7. *original and translation ceased publication;*
8. *commencing 1965, the Soviet journal was split into "Izvestiya Akademii Nauk SSSR. Seriya Fiziki Atmosfery i Okeana", and "Izvestiya Akademii Nauk SSSR. Seriya Fiziki Zemli" (both of these are being translated).*

108 IZVESTIYA AKADEMII NAUK SSSR. SERIYA GEOLOGICHESKAYA
(Izvestiya of the Academy of Sciences of the USSR. Geologic Series)
Trans. covers 1958–1961, vols. 23–26.
Order from: KP
1.a; 2.a; 3.c; 4.a; 5.a; 6.a;
7. translation ceased publication;
8. commencing 1962, v. 27+, only selected articles are translated (in "International Geology Review"), but translation of all articles published in the Soviet journal are available from AGI.

109 IZVESTIYA AKADEMII NAUK SSSR. SERIYA KHIMICHESKAYA
(Bulletin of the Academy of Sciences, USSR. Division of Chemical Science)
Trans. begins with 1952, v. 16+
Monthly
Order from: PLPC
1.a; 2.a; 3.c; 4.a; 5.a; 6.a;
7. 12 mos.;
8. prior to 1963, no. 7, the Soviet journal was titled "Izvestiya Akademii Nauk SSSR. Otdelenie Khimicheskaya Nauk".

110 IZVESTIYA AKADEMII NAUK SSSR. SERIYA MATEMATICHESKAYA
(Transactions of the Academy of Science USSR. Mathematics Series)
Trans. begins with 1967, v. 31+
Bimonthly
Order from: AMS
1.a; 2.a; 3.a; 4.a; 5.a; 6. ;
7. 12 mos.

111 IZVESTIYA AKADEMII NAUK SSSR. SERIYA NEORGANICHESKIE MATERIALY
(Inorganic Materials)
Trans. begins with 1965, v. 1+
Monthly
Order from: PLPC
1.a; 2.a; 3.a; 4.a; 5.a; 6.a;
7. 6 mos.

112 IZVESTIYA AKADEMII NAUK SSSR. TEKHNICHESKAYA KIBERNETIKA
(Engineering Cybernetics)
Trans. begins with 1963, no. 1+
Bimonthly
Order from: IEEE or ST
1.a; 2.a; 3.a; 4.a; 5.b; 6.b;
7. 7 mos.;
8. another translation, "Technical Cybernetics", is available from NTIS; the Soviet journal changed title with 1964, no. 4, from "Izvestiya Akademii Nauk SSSR. Otdelenie Tekhnicheskikh Nauk. Tekhnicheskaya Kibernetika".

113 IZVESTIYA SIBIRSKOGO OTDELENIYA AKADEMII NAUK SSSR. SERIYA KHIMICHESKIKH NAUK
(Siberian Chemistry Journal)
Trans. covers 1966–1970, vols. 1–5.
Bimonthly
Order from: PLPC
1.a; 2.a; 3.a; 4.a; 5.a; 6.a;
7. translation ceased publication.

114 IZVESTIYA VYSSHIKH UCHEBNYKH ZAVEDENII. AVIATSIONNAYA TEKHNIKA
(Soviet Aeronautics)
Trans. begins with 1966, v. 9+
Quarterly
Order from: APr
1.a; 2.a; 3.a; 4.a; 5.a; 6.a;
7. the present publisher of this translation is making every effort to reduce previously existing translation delays and it is deemed advisable not to cite any specific "lag time" in this edition of the GUIDE.
8. v. 6, nos. 1–3, 1963, titled "Aviation Engineering", are available from NTIS.

IZVESTIYA VYSSHIKH UCHEBNYKH ZAVEDENII. CHERNAYA METALLURGIYA
see "Steel in the USSR" (Part II)

115 IZVESTIYA VYSSHIKH UCHEBNYKH ZAVEDENII. FIZIKA
(Soviet Physics Journal)
Trans. begins with 1965, v. 7+
Bimonthly, vols. 8–9; Monthly v. 10+
Order from: PLPC
1.a; 2.a; 3.a; 4.a; 5.a; 6.a;
[continued next page]

EXPLANATION FOR PART I ENTRIES

** *Indicates that the title has been announced for publication in translation, but has not been so published as of the date of this study (June 1, 1972). Consult publisher for further information.*

Trans. begins — refers to year and volume of the original publication. Unless otherwise indicated, translated volume commences with issue number 1.

Trans. covers — refers to volume (or volumes) translated before translation ceased publication.

OP — following publisher code indicates back issues no longer available from the publisher.

<u>Explanation of Coded data:</u>

1. a — *Cover-to-cover translation*
   b — *Selected articles translated*

2. a — *Full translations of all articles*
   b — *Some articles fully translated, others abridged or abstracted*
   c — *Selected articles abridged or abstracted*
   d — *Articles translated are fully translated, others merely cited or listed*

3. a — *Translation carries original language numbers*
   b — *Translation carries other numbers*
   c — *Translation carries no numbers*

4. a — *Translation carries original language dates*
   b — *Translation carries other dates*
   c — *Translation carries no dates*

5. a — *Translation identifies original pagination*
   b — *Translation does not identify original pagination*

6. a — *Translation has volume index or collective table of contents*
   b — *Translation has no volume index or collective table of contents*

7. *"Lag time" between month of issue of original publication and month of issue of translated counter-part (there is a possible variation of one month because of mailing practices both of the original and the translation)*

8. *Other information (specified)*

9. *Translation exists but has not been located for verification and evaluation*

7. *the present publisher of this translation is making every effort to reduce previously existing translation delays and it is deemed advisable not to cite any specific "lag time" in this edition of the GUIDE.*

116 IZVESTIYA VYSSHIKH UCHEBNYKH ZAVEDENII. GEODEZIYA I AEROFOTOS'EMKA
(Geodesy and Aerophotography)
Trans. begins with 1962, v. 7+
Bimonthly
Order from: AGU
1.a; 2.a; 3.a; 4.a; 5.a; 6.a;
7. 25 mos.

117 IZVESTIYA VYSSHIKH UCHEBNYKH ZAVEDENII. PRIBOROSTROENIE
(Izvestiya VUZOV. Instrument Building)
Trans. covers 1962–1966, vols. 5–9.
Bimonthly
Order from: NTIS
8. vols. 5–6, 1962–1963, were translated as "News of Higher Educational Institutions: Instrument Building".
9.

118 IZVESTIYA VYSSHIKH UCHEBNYKH ZAVEDENII. RADIOFIZIKA
(Radiophysics and Quantum Electronics)
Trans. begins with 1959, v. 2+
Monthly
Order from: PLPC
1.a; 2.a; 3.a; 4.a; 5.a; 6.a;
7. the present publisher of this translation is making every effort to reduce previously existing translation delays and it is deemed advisable not to cite any specific "lag time" in this edition of the GUIDE.
8. vols. 2–6, 1959–1963, were translated as "News of Higher Educational Institutions: Radiophysics Series", and vols. 7–8, no. 2, 1964–1965, were translated as "Izvestiya VUZOV. Radiophysics", both available from NTIS; the PLPC translation commences with 1965, v. 8+, and was titled "Soviet Radiophysics" for vols. 8 and 9.

119 IZVESTIYA VYSSHIKH UCHEBNYKH ZAVEDENII. RADIOTEKHNIKA
(Soviet Radio Engineering)
Trans. begins with 1959, v. 2+
Bimonthly
Order from: FP-OP
[continued next column]

1.a; 2.a; 3.a; 4.a; 5.a; 6.a;
7. translation ceased publication, termination date unknown;
8. vols. 2–6, 1959–1963, were translated as "News of Higher Educational Institutions: Radio Engineering Series", and vols. 7–8, no. 2, 1964–1965, were translated as "Izvestiya VUZOV. Radio Engineering", both available from NTIS; the FP translation commences with 1965, v. 8+.

120 IZVESTIYA VYSSHIKH UCHEBNYKH ZAVEDENII. TEKHNOLOGIYA TEKSTIL'NOI PROMYSHLENNOST'
(Technology of the Textile Industry, USSR)
Trans. begins with 1960, v. 4+
Bimonthly
Order from: TI
1.a; 2.a; 3.a; 4.a; 5.b; 6.a;
7. 10 mos.

121 IZVESTIYA VYSSHIKH UCHEBNYKH ZAVEDENII. TSVETNAYA METALLURGIYA
(Nonferrous Metallurgy)
Trans. covers 1963, v. 6.
Bimonthly
Order from: NTIS
1.a; 2.a; 3.c; 4.a; 5.a; 6.b;
7. translation ceased publication.

JOURNAL OF THE ELECTROCHEMICAL SOCIETY OF JAPAN see
"Denki Kagaku"

JOURNAL OF THE ELECTRONIC COMMUNICATION ENGINEERS OF JAPAN see
"Denki Tsushin Gakkai Zasshi" and "Denshi Tsushin Gakkai Ronbunshi"

122 KAGAKU KOGAKU
(Chemical Engineering, Japan –[Abridged Edition in English])
Semi-annual
Order from: MAR
1.b; 2.b; 3.b; 4.b; 5.a; 6.b;
7. 9 mos.;
8. published by the Society of Chemical Engineers, Japan.

123 KAUCHUK I REZINA
(Soviet Rubber Technology)
Trans. begins with 1959, v. 18+
Monthly
[continued next page]

EXPLANATION FOR PART I ENTRIES

** *Indicates that the title has been announced for publication in translation, but has not been so published as of the date of this study (June 1, 1972). Consult publisher for further information.*

Trans. begins —— refers to year and volume of the original publication. Unless otherwise indicated, translated volume commences with issue number 1.

Trans. covers —— refers to volume (or volumes) translated before translation ceased publication.

OP —— following publisher code indicates back issues no longer available from the publisher.

<u>*Explanation of Coded data:*</u>

1. *a — Cover-to-cover translation*
   b — Selected articles translated

2. *a — Full translations of all articles*
   b — Some articles fully translated, others abridged or abstracted
   c — Selected articles abridged or abstracted
   d — Articles translated are fully translated, others merely cited or listed

3. *a — Translation carries original language numbers*
   b — Translation carries other numbers
   c — Translation carries no numbers

4. *a — Translation carries original language dates*
   b — Translation carries other dates
   c — Translation carries no dates

5. *a — Translation identifies original pagination*
   b — Translation does not identify original pagination

6. *a — Translation has volume index or collective table of contents*
   b — Translation has no volume index or collective table of contents

7. *"Lag time" between month of issue of original publication and month of issue of translated counter-part (there is a possible variation of one month because of mailing practices both of the original and the translation)*

8. *Other information (specified)*

9. *Translation exists but has not been located for verification and evaluation*

Order from: PPC or Ma & S
1.a; 2.b; 3.a; 4.a; 5.a; 6.b;
7. 12 mos.

124 KHIMICHESKAYA PROMYSHLENNOST'
(The Soviet Chemical Industry)
Trans. begins with 1969, no. 7+
Monthly
Order from: McE
1.a; 2.a; 3.a; 4.a; 5.a; 6. ;
7. 2 mos.

125 KHIMICHESKIE VOLOKNA
(Fibre Chemistry)
Trans. begins with 1969, v. 1+
Bimonthly
Order from: PLPC
1.a; 2.a; 3.a; 4.a; 5.a; 6.a;
7. 6 mos.

126 KHIMICHESKOE I NEFTYANOE MASHINO-STROENIE
(Chemical and Petroleum Engineering)
Trans. begins with 1965, v. 1+
Monthly
Order from: PLPC
1.a; 2.a; 3.a; 4.a; 5.a; 6.a;
7. 6 mos.;
8. Soviet publication carried title "Khimicheskoe Mashinostroenie" until 1966.

127 KHIMICHESKOE MASHINOSTROENIE
(Chemical Engineering)
Trans. covers 1964–1965.
Order from: PLPC
1.a; 2.a; 3.a; 4.a; 5.a; 6.a;
7. original and translation ceased publication, see "Khimicheskoe i Neftyanoe Mashinostroenie" (Chemical and Petroleum Engineering) for successor publication.

128 KHIMIKO-FARMATSEVTICHESKII ZHURNAL
(Pharmaceutical Chemistry Journal)
Trans. begins with 1967, v. 1+
Monthly
Order from: PLPC
1.a; 2.a; 3.a; 4.a; 5.a; 6.a;
7. 6 mos.

129 KHIMIYA GETEROTSIKLICHESKIKH SOEDINENII
(Chemistry of Heterocyclic Compounds)
Trans. begins with 1965, v. 1+
Bimonthly, vols. 1–5 (1965–1969);
Monthly, v. 6+
Order from: PLPC
1.a; 2.a; 3.a; 4.a; 5.a; 6.a;
7. the present publisher of this translation is making every effort to reduce previously existing translation delays and it is deemed advisable not to cite any specific "lag time" in this edition of the GUIDE.

130 KHIMIYA I TEKHNOLOGIYA TOPLIV I MASEL
(Chemistry and Technology of Fuels and Oils)
Trans. begins with 1965, v. 10+
Monthly
Order from: PLPC
1.a; 2.a; 3.a; 4.a; 5.a; 6.a;
7. 6 mos.

131 KHIMIYA PRIRODNYKH SOEDINENII
(Chemistry of Natural Compounds)
Trans. begins with 1965, v. 1+
Bimonthly
Order from: PLPC
1.a; 2.a; 3.a; 4.a; 5.a; 6.a;
7. the present publisher of this translation is making every effort to reduce previously existing translation delays and it is deemed advisable not to cite any specific "lag time" in this edition of the GUIDE.

132 KHIMIYA VYSOKIKH ENERGII
(High Energy Chemistry)
Trans. begins with 1967, v. 1+
Bimonthly
Order from PLPC
1.a; 2.a; 3.a; 4.a; 5.a; 6. ;
7. 6 mos.

133 KIBERNETIKA
(Cybernetics)
Trans. begins with 1965, v. 1+
Bimonthly
Order from: PLPC
1.a; 2.a; 3.a; 4.a; 5.a; 6.a;
[continued next page]

EXPLANATION FOR PART I ENTRIES

** *Indicates that the title has been announced for publication in translation, but has not been so published as of the date of this study (June 1, 1972). Consult publisher for further information.*

Trans. begins — refers to year and volume of the original publication. Unless otherwise indicated, translated volume commences with issue number 1.

Trans. covers — refers to volume (or volumes) translated before translation ceased publication.

OP — following publisher code indicates back issues no longer available from the publisher.

<u>Explanation of Coded data:</u>

1. a — Cover-to-cover translation
   b — Selected articles translated

2. a — Full translations of all articles
   b — Some articles fully translated, others abridged or abstracted
   c — Selected articles abridged or abstracted
   d — Articles translated are fully translated, others merely cited or listed

3. a — Translation carries original language numbers
   b — Translation carries other numbers
   c — Translation carries no numbers

4. a — Translation carries original language dates
   b — Translation carries other dates
   c — Translation carries no dates

5. a — Translation identifies original pagination
   b — Translation does not identify original pagination

6. a — Translation has volume index or collective table of contents
   b — Translation has no volume index or collective table of contents

7. "Lag time" between month of issue of original publication and month of issue of translated counter-part (there is a possible variation of one month because of mailing practices both of the original and the translation)

8. Other information (specified)

9. Translation exists but has not been located for verification and evaluation

7. the present publisher of this translation is making every effort to reduce previously existing translation delays and it is deemed advisable not to cite any specific "lag time" in this edition of the GUIDE.
8. another translation, with an identical title, is published by NTIS.

134 KINETIKA I KATALIZ
(Kinetics and Catalysis)
Trans. begins with 1960, v. 1+
Bimonthly
Order from: PLPC
1.a; 2.a; 3.a; 4.a; 5.a; 6.a;
7. 6 mos.

135 KLINICHESKAYA MEDITSINA
(Clinical Medicine)
Trans. covers 1966, v. 44, nos. 1–11.
Monthly
Order from: NTIS
9.

136 KOKS I KHIMIYA
(Coke and Chemistry, USSR)
Trans. begins with 1959, v. 16, no. 8+
Monthly
Order from: CTRA
1.b; 2.d; 3.c; 4.a; 5.b; 6.b;
7. 6 mos.

137 KOLLOIDNYI ZHURNAL
(Colloid Journal of the USSR)
Trans. begins with 1952, v. 14+
Bimonthly
Order from: PLPC
1.a; 2.a; 3.a; 4.a; 5.a; 6.a;
7. 6 mos.

138 KOSMICHESKAYA BIOLOGIYA I MEDITSINA
(Space Biology and Medicine)
Trans. begins with 1967, v. 1+
Order from: NTIS
1.a; 2.a; 3.a; 4.a; 5.a; 6.a;
7. 4 mos.;
8. vols. 1–4, 1967–1970, were published by PLPC under the title "Environmental Space Sciences".

139 KOSMICHESKIE ISSLEDOVANIYA
(Cosmic Research)
Trans. begins with 1963, v. 1+
Bimonthly
[continued next column]

Order from: PLPC
1.a; 2.a; 3.a; 4.a; 5.a; 6.a;
7. 6 mos.;
8. another translation is available from NTIS.

140 KRISTALLOGRAFIYA
(Soviet Physics—Crystallography)
Trans. begins with 1956, v. 1+
Bimonthly
Order from: AIP
1.a; 2.a; 3.a; 4.a; 5.a; 6.a;
7. 6 mos.

141 KVANTOVAYA ELEKTRONIKA
(Soviet Journal of Quantum Electronics)
Trans. begins with 1971, v. 1+
Bimonthly
Order from: AIP
1.a; 2.a; 3.a; 4.a; 5.a; 6.a;
7. 6 mos.

LI HSUEH HSUEH PAO see
"Acta Mechanica Sinica"

142 LIJECNICKI VJESNIK
(Medical Journal)
Trans. begins with 1962, v. 84+
Monthly
Order from: NTIS
1.a; 2.a; 3.a; 4.a; 5.a; 6.b;
7. 12 mos.

143 LITEINOE PROIZVODSTVO
(Russian Castings Production)
Trans. begins with 1961, v. 12+
Monthly
Order from: BCIRA
1.a; 2.a; 3.a; 4.a; 5.b; 6.a;
7. 12 mos.;
8. index issued approximately 12 mos. late; another translation, "Giessereiwessen", available from DBEI.

144 LITOLOGIYA I POLEZNYE ISKOPAEMYE
(Lithology and Mineral Resources)
Trans. begins with 1966+
Bimonthly
Order from: PLPC
1.a; 2.a; 3.a; 4.a; 5.a; 6.a;
7. 6 mos.

EXPLANATION FOR PART I ENTRIES

** *Indicates that the title has been announced for publication in translation, but has not been so published as of the date of this study (June 1, 1972). Consult publisher for further information.*

Trans. begins — refers to year and volume of the original publication. Unless otherwise indicated, translated volume commences with issue number 1.

Trans. covers — refers to volume (or volumes) translated before translation ceased publication.

OP — following publisher code indicates back issues no longer available from the publisher.

Explanation of Coded data:

1. *a — Cover-to-cover translation*
   b — Selected articles translated

2. *a — Full translations of all articles*
   b — Some articles fully translated, others abridged or abstracted
   c — Selected articles abridged or abstracted
   d — Articles translated are fully translated, others merely cited or listed

3. *a — Translation carries original language numbers*
   b — Translation carries other numbers
   c — Translation carries no numbers

4. *a — Translation carries original language dates*
   b — Translation carries other dates
   c — Translation carries no dates

5. *a — Translation identifies original pagination*
   b — Translation does not identify original pagination

6. *a — Translation has volume index or collective table of contents*
   b — Translation has no volume index or collective table of contents

7. *"Lag time" between month of issue of original publication and month of issue of translated counter-part (there is a possible variation of one month because of mailing practices both of the original and the translation)*

8. *Other information (specified)*

9. *Translation exists but has not been located for verification and evaluation*

145 MAGNITNAYA GIDRODINAMIKA
(Magnetohydrodyamics)
Trans. begins with 1965, v. 1+
Quarterly
Order from: PLPC
1.a; 2.a; 3.a; 4.a; 5.a; 6.a;
7. the present publisher of this translation is making every effort to reduce previously existing translation delays and it is deemed advisable not to cite any specific "lag time" in this edition of the GUIDE.

146 MASHINOVEDENIE
(Machine Science Abstracts)
Trans. begins with 1965, v. 1+
Annual vols. 1965–1966,
Bimonthly 1967+
Order from: SIC
8. publication carries abstracts of every article with illustrations and full bibliography;
9.

147 MATEMATICHESKII SBORNIK
(Mathematics of the USSR–Sbornik)
Trans. begins with 1967, v. 1 (New Series)+
Monthly
Order from: AMS
1.a; 2.a; 3.b; 4.a; 5.a; 6.a;
7. 12 mos.

148 MATEMATICHESKII ZAMETKI
(Mathematical Notes)
Trans. begins with 1967, v. 1+
Bimonthly
Order from: PLPC
1.a; 2.a; 3.a; 4.a; 5.a; 6.a;
7. 6 mos.

149 MEDITSINSKAYA PROMYSHLENNOST' SSSR
(Medical Industry of the USSR)
Trans. covers 1966, v. 20, nos. 1–9.
Monthly
Order from: NTIS
9.

150 MEDITSINSKAYA RADIOLOGIYA
(Medical Radiology)
Trans. covers 1959–1962, vols. 4–7.
Monthly
Order from: NTIS
7. translation ceased publication;
9.

151 MEDITSINSKAYA TEKHNIKA
(Biomedical Engineering)
Trans. begins with 1967, v. 1+
Bimonthly
Order from: PLPC
1.a; 2.a; 3.a; 4.a; 5.a; 6.a;
7. 7 mos.

152 MEDITSINSKII REFERATIVNYI ZHURNAL
(Abstracts of Soviet Medicine)
Trans. covers 1957–1961, vols. 1–5.
Monthly
Order from: EMF
7. translation ceased publication;
9.

153 MEDYCYNA DOSWIADCZALNA I MIKROBIOLOGIA
(Experimental Medicine and Microbiology)
Trans. begins with 1963, v. 15+
Quarterly
Order from: NTIS
1.a; 2.a; 3.a; 4.a; 5.b; 6.a;
7. 9 mos.

154 MEKHANIKA POLIMEROV
(Polymer Mechanics)
Trans. begins with 1965, v. 1+
Bimonthly
Order from: PLPC
1.a; 2.a; 3.a; 4.a; 5.a; 6.a;
7. the present publisher of this translation is making every effort to reduce previously existing translation delays and it is deemed advisable not to cite any specific "lag time" in this edition of the GUIDE.

METALLOVEDENIE I OBRABOTKA METALLOV see
"Metallovedenie i Termicheskaya Obrabotka Metallov"

155 METALLOVEDENIE I TERMICHESKAYA OBRABOTKA METALLOV
(Metal Science and Heat Treatment)
Trans. begins with 1958+
Bimonthly
Order from: PLPC
1.a; 2.a; 3.a; 4.a; 5.a; 6.a;
7. 6 mos.;
8. each translated issue covers two original issues; 1958 Russian title was "Metallovedenie i Obrabotka Metallov".

EXPLANATION FOR PART I ENTRIES

** *Indicates that the title has been announced for publication in translation, but has not been so published as of the date of this study (June 1, 1972). Consult publisher for further information.*

Trans. begins — refers to year and volume of the original publication. Unless otherwise indicated, translated volume commences with issue number 1.

Trans. covers — refers to volume (or volumes) translated before translation ceased publication.

OP — following publisher code indicates back issues no longer available from the publisher.

Explanation of Coded data:

1. a — Cover-to-cover translation
   b — Selected articles translated

2. a — Full translations of all articles
   b — Some articles fully translated, others abridged or abstracted
   c — Selected articles abridged or abstracted
   d — Articles translated are fully translated, others merely cited or listed

3. a — Translation carries original language numbers
   b — Translation carries other numbers
   c — Translation carries no numbers

4. a — Translation carries original language dates
   b — Translation carries other dates
   c — Translation carries no dates

5. a — Translation identifies original pagination
   b — Translation does not identify original pagination

6. a — Translation has volume index or collective table of contents
   b — Translation has no volume index or collective table of contents

7. "Lag time" between month of issue of original publication and month of issue of translated counter-part (there is a possible variation of one month because of mailing practices both of the original and the translation)

8. Other information (specified)

9. Translation exists but has not been located for verification and evaluation

156 METALLURG
(Metallurgist)
Trans. begins with 1957, v. 2+
Bimonthly
Order from: PLPC
1.a; 2.a; 3.a; 4.a; 5.a; 6.a;
7. 6 mos.

157 METEORITIKA
(Meteoritica)
Trans. begins with 1963, v. 23+
Order from: STR
1.a; 2.a; 3.a; 4.a; 5.b; 6.a;
7. 3 years (only one issue published to date);
8. each volume is a single issue.

158 MIKROBIOLOGIYA
(Microbiology)
Trans. begins with 1957, v. 26+
Bimonthly
Order from: PLPC
1.a; 2.a; 3.a; 4.a; 5.a; 6.a;
7. 6 mos.;
8. vols. 26−30, 1957−1961, are available from NTIS; the PLPC translation begins with 1962, v. 31+.

159 MIKROELEKTRONIKA**
(Microelectronics)
Trans. to begin with Russian 1972, no. 1+
Order from: PLPC

160 MOLEKULYARNAYA BIOLOGIYA
(Molecular Biology)
Trans. begins with 1967, v. 1+
Bimonthly
Order from: PLPC
1.a; 2.a; 3.a; 4.a; 5.a; 6. ;
7. 6 mos.

NAUCHNO-TEKHNICHESKII PROBLEMY GORENIYA I VZRYVA see
"Fizika Goreniya i Vzryva"

161 NEFTEKHIMIYA
(Petroleum Chemistry, USSR)
Trans. begins with 1961, v. 1+
Quarterly
Order from: PP
1.b; 2.d; 3.a; 4.b; 5.a; 6.a;
7. 12 mos.

162 NEIROFIZIOLOGIYA
(Neurophysiology)
Trans. begins with 1969, v. 1+
Bimonthly
Order from: PLPC
1.a; 2.a; 3.a; 4.a; 5.a; 6.a;
7. 8 mos.

163 NUKLEONIKA
(Nucleonics)
Trans. covers 1957−1962, vol. 2−7, and begins again with 1965, v. 10+
Montbly
Order from: NTIS
1.a; 2.a; 3.a; 4.a; 5.a; 6.a;
7. 12 mos.;
8. vols. 2−5 carry only selected articles, v. 6+ are cover-to-cover translations, vol. 8−9, 1963−1964, are in the process of being translated.

164 OGNEUPORY
(Refractories)
Trans. begins with 1960, v. 25+
Bimonthly
Order from: PLPC
1.a; 2.a; 3.a; 4.a; 5.a; 6.b;
7. 6 mos.;
8. 1960−1961 issues published by Acta Metallurgica, PLPC translation commences with 1962, no. 1+.

165 OKEANOLOGIYA
(Oceanology)
Trans. begins with 1965, v. 5+
Bimonthly
Order from: AGU
1.a; 2.a; 3.a; 4.a; 5.a; 6.a;
7. 8 mos.

166 ONTOGENEZ
(Soviet Journal of Developmental Biology)
Trans. begins with 1970, v. 1+
Bimonthly
Order from: PLPC
1.a; 2.a; 3.a; 4.a; 5.a; 6.a;
7. 8 mos.

167 OPTIKA I SPEKTROSKOPIYA
(Optics and Spectroscopy)
Trans. begins with 1959, v. 6+
Montbly
Order from: OSA
[continued next page]

EXPLANATION FOR PART I ENTRIES

** *Indicates that the title has been announced for publication in translation, but has not been so published as of the date of this study (June 1, 1972). Consult publisher for further information.*

Trans. begins — refers to year and volume of the original publication. Unless otherwise indicated, translated volume commences with issue number 1.

Trans. covers — refers to volume (or volumes) translated before translation ceased publication.

OP — following publisher code indicates back issues no longer available from the publisher.

Explanation of Coded data:

1. a — *Cover-to-cover translation*
   b — *Selected articles translated*

2. a — *Full translations of all articles*
   b — *Some articles fully translated, others abridged or abstracted*
   c — *Selected articles abridged or abstracted*
   d — *Articles translated are fully translated, others merely cited or listed*

3. a — *Translation carries original language numbers*
   b — *Translation carries other numbers*
   c — *Translation carries no numbers*

4. a — *Translation carries original language dates*
   b — *Translation carries other dates*
   c — *Translation carries no dates*

5. a — *Translation identifies original pagination*
   b — *Translation does not identify original pagination*

6. a — *Translation has volume index or collective table of contents*
   b — *Translation has no volume index or collective table of contents*

7. *"Lag time" between month of issue of original publication and month of issue of translated counter-part (there is a possible variation of one month because of mailing practices both of the original and the translation)*

8. *Other information (specified)*

9. *Translation exists but has not been located for verification and evaluation*

1.a; 2.a; 3.b; 4.a; 5.a; 6.a;
7. 5 mos.;
8. a cummulative index to vols. 1–10 is available from OSA, this lists individual papers translated from vols. 1–5 and sources of their availability.

168 **OPTIKO-MEKHANICHESKAYA PROMYSH-LENNOST'**
(Soviet Journal of Optical Technology)
Trans. begins with 1966, v. 33+
Bimonthly
Order from: AIP
1.a; 2.a; 3.a; 4.a; 5.a; 6.a;
7. 9 mos.

169 **OSNOVANIYA FUNDAMENTY I MEKHANIKA GRUNTOV**
(Soil Mechanics and Foundation Engineering)
Trans. begins with 1965, v. 7+
Bimonthly
Order from: PLPC
1.a; 2.a; 3.a; 4.a; 5.a; 6.a;
7. 6 mos.

170 **OTKRITIYA, IZOBRETENIYA, PROMYSHLENNYE OBRAZTSI, TOVARNYE ZNAKI**
(Soviet Inventions and Patents)
Trans. begins with 1971, no. 1+
36 issues per yr.
Order from: SIC
9.

171 **PALEONTOLOGICHESKII ZHURNAL**
(Paleontological Journal)
Trans. begins with 1962, v. 1+
Quarterly
Order from: AGI
8. prior to 1962, while the entire journal was translated, only selected papers were published in "International Geology Review", though articles not translated in IGR are available from IGR at a nominal cost; commencing 1967+, the entire journal is translated cover-to cover;
9.

172 **PLASTICHESKIE MASSY**
(Soviet Plastics)
Trans. begins with 1960, v. 7+
Monthly
Order from: RAPRA
1.a; 2.a; 3.c; 4.a; 5.a; 6.b;
7. 13 mos.

173 **POCHVOVEDENIE**
(Soviet Soil Science)
Trans. begins with 1958+
Bimonthly
Order from: SSSA or ST
1.a; 2.a; 3.a; 4.a; 5.a; 6.a;
7. 7 mos.;
8. 1958–1961 issues available from NTIS; 1962–1963 issues available from ST; 1962, no. 13 and 1964, no. 13 issues of this journal are supplements containing only selected articles (titled "Doklady Soil Science Sections").

174 **POLSKI PRZEGLAD RADIOLOGII I MEDYCYNY NUKLEARNEJ**
(Polish Review of Radiology and Nuclear Medicine)
Trans. begins with 1963, v. 27+
Bimonthly
Order from: NTIS
1.a; 2.a; 3.a; 4.a; 5.b; 6.a;
7. 12 mos.

175 **POROSHKOVAYA METALLURGIYA**
(Soviet Powder Metallurgy [and Metal Ceramics])
Trans. begins with 1962, v. 2+
Bimonthly 1962–1964,
Monthly 1965+
Order from: PLPC
1.a; 2.a; 3.a; 4.a; 5.a; 6.a;
7. 6 mos.

176 **PRIBOROSTROENIE**
(Instrument Construction)
Trans. covers 1959–1966, vols. 4–8.
Monthly
Order from: TF
1.a; 2.a; 3.a; 4.a; 5.a; 6.b;
8. original and translation changed titles commencing 1967 (see "PRIBORY I SISTEMY UPRAVLENIYA").

177 **PRIBORY I SISTEMY UPRAVLENIYA**
(Soviet Journal of Instrumentation and Control)
Trans. begins with 1967, no. 1+
Monthly
Order from: STE
1.a; 2.a; 3.a; 4.a; 5.a; 6.b;
7. 12 mos.;
8. formerly titled "Priborostroenie" and translated as "Instrument Construction".

EXPLANATION FOR PART I ENTRIES

** *Indicates that the title has been announced for publication in translation, but has not been so published as of the date of this study (June 1, 1972). Consult publisher for further information.*

Trans. begins — refers to year and volume of the original publication. Unless otherwise indicated, translated volume commences with issue number 1.

Trans. covers — refers to volume (or volumes) translated before translation ceased publication.

OP — following publisher code indicates back issues no longer available from the publisher.

Explanation of Coded data:

1. *a — Cover-to-cover translation*
   b — Selected articles translated

2. *a — Full translations of all articles*
   b — Some articles fully translated, others abridged or abstracted
   c — Selected articles abridged or abstracted
   d — Articles translated are fully translated, others merely cited or listed

3. *a — Translation carries original language numbers*
   b — Translation carries other numbers
   c — Translation carries no numbers

4. *a — Translation carries original language dates*
   b — Translation carries other dates
   c — Translation carries no dates

5. *a — Translation identifies original pagination*
   b — Translation does not identify original pagination

6. *a — Translation has volume index or collective table of contents*
   b — Translation has no volume index or collective table of contents

7. *"Lag time" between month of issue of original publication and month of issue of translated counter-part (there is a possible variation of one month because of mailing practices both of the original and the translation)*

8. *Other information (specified)*

9. *Translation exists but has not been located for verification and evaluation*

178 PRIBORY I TEKHNIKA EKSPERIMENTA
(Instruments and Experimental Techniques)
Trans. begins with 1958, v. 3+
Monthly
Order from: PLPC
1.a; 2.a; 3.a; 4.a; 5.a; 6.a;
7. 6 mos.

179 PRIKLADNAYA BIOKHIMIYA I MIKROBIOLOGIYA
(Applied Biochemistry and Microbiology)
Trans. begins with 1965, v. 1+
Bimonthly
Order from: PLPC
1.a; 2.a; 3.a; 4.a; 5.a; 6.a;
7. the present publisher of this translation is making every effort to reduce previously existing translation delays and it is deemed advisable not to cite any specific "lag time" in this edition of the GUIDE.

180 PRIKLADNAYA GEOFIZIKA
(Exploration Geophysics)
Trans. covers 1968–1969, vols. 47–51.
Annual
Order from: PLPC

181 PRIKLADNAYA MATEMATIKA I MEKHANIKA
(Journal of Applied Mathematics and Mechanics)
Trans. begins with 1958, v. 22+
Bimonthly
Order from: PP
1.a; 2.a; 3.a; 4.a; 5.a; 6.a;
7. 6 mos.

182 PRIKLADNAYA MEKHANIKA
(Soviet Applied Mechanics)
Trans. begins with 1966, v. 2+
Monthly
Order from: PLPC
1.a; 2.a; 3.a; 4.a; 5.a; 6.a;
7. the present publisher of this translation is making every effort to reduce previously existing translation delays and it is deemed advisable not to cite any specific "lag time" in this edition of the GUIDE.

183 PROBLEMY GEMATOLOGII I PERELIVANIYA KROVI
(Problems of Hematology and Blood Transfusion)
Trans. covers 1957–1961, vols. 2–6.
Monthly
Order from: PP
1.a; 2.a; 3.a; 4.a; 5.a; 6.a;
7. translation ceased publication.

184 PROBLEMY KIBERNETIKI
(Problems of Cybernetics)
Trans. covers 1959–1963, vols. 4–8.
Irregular
Order from: PP
1.a; 2.a; 3.a; 4.a; 5.a; 6.a;
8. another translation is available from NTIS covering 1962, no. 7– 1967, no. 18; another trnaslation, "Probleme der Kybernetik", 1959+, is available from the Zentralinstitut fur Automatisierung, Berlin.

185 PROBLEMY PEREDACHI INFORMATSII
(Problems of Information Transmission)
Trans. begins with 1965, v. 1+
Quarterly
Order from: PLPC
1.a; 2.a; 3.a; 4.a; 5.a; 6.a;
7. the present publisher of this translation is making every effort to reduce previously existing translation delays and it is deemed advisable not to cite any specific "lag time" in this edition of the GUIDE.
8. another translation is available from NTIS covering 1966–1967, vols. 2–3, no. 4.

186 PROBLEMY PROCHNOSTI
(Strength of Materials)
Trans. begins with 1969, v. 1+
Monthly
Order from: PLPC
1.a; 2.a; 3.a; 4.a; 5.a; 6.a;
7. 6 mos.

187 PROBLEMY SEVERA
(Problems of the North)
Trans. begins with 1958+
Irregular
Order from: NRC
1.a; 2.a; 3.a; 4.a; 5.b; 6.b;
7. 18 mos.

188 PRZEGLAD EPIDEMIOLOGICZNY
(Epidemiological Review)
Trans. begins with 1963, v. 17+
Semiannual
Order from: NTIS
1.a; 2.a; 3.a; 4.a; 5.b; 6.a;
7. 10 mos.

189 PRZEMYSL CHEMICZNY**
(Polish Chemical Industry)
Trans. to begin with 1968
Monthly
[continued next page]

EXPLANATION FOR PART I ENTRIES

** *Indicates that the title has been announced for publication in translation, but has not been so published as of the date of this study (June 1, 1972). Consult publisher for further information.*

Trans. begins — refers to year and volume of the original publication. Unless otherwise indicated, translated volume commences with issue number 1.

Trans. covers — refers to volume (or volumes) translated before translation ceased publication.

OP — following publisher code indicates back issues no longer available from the publisher.

Explanation of Coded data:

1. a — *Cover-to-cover translation*
   b — *Selected articles translated*

2. a — *Full translations of all articles*
   b — *Some articles fully translated, others abridged or abstracted*
   c — *Selected articles abridged or abstracted*
   d — *Articles translated are fully translated, others merely cited or listed*

3. a — *Translation carries original language numbers*
   b — *Translation carries other numbers*
   c — *Translation carries no numbers*

4. a — *Translation carries original language dates*
   b — *Translation carries other dates*
   c — *Translation carries no dates*

5. a — *Translation identifies original pagination*
   b — *Translation does not identify original pagination*

6. a — *Translation has volume index or collective table of contents*
   b — *Translation has no volume index or collective table of contents*

7. *"Lag time" between month of issue of original publication and month of issue of translated counter-part (there is a possible variation of one month because of mailing practices both of the original and the translation)*

8. *Other information (specified)*

9. *Translation exists but has not been located for verification and evaluation*

Order from: SI
8. translations of original papers will be complete; remaining material will be abstracted with translations of the full papers available upon request.

190 RADIOBIOLOGIYA
(Radiobiology)
Trans. begins with 1961, v. 1+
Bimonthly
Order from: NTIS
1.a; 2.a; 3.a; 4.a; 5.a; 6.b;
7. 12 mos.;
8. vol. 8, nos. 4–6 not translated.

191 RADIOKHIMIYA
(Soviet Radiochemistry)
Trans. begins with 1959, v. 1+
Bimonthly
Order from: PLPC
1.a; 2.a; 3.a; 4.a; 5.a; 6.a;
7. 6 mos.;
8. v. 1, 1959, is available from PP as "Radiochemistry"; vols. 2–3, 1960–1961, were issued by NTIS as two joint issues; PLPC became the publisher commencing 1962, v. 4+; another translation, "Radiochemie, 1965–1970, vols. 6–9, no. 6, is available from CNRS.

192 RADIOTEKHNIKA
(Radio Engineering)
Trans. covers 1957–1963, vols. 12–16.
Monthly
Order from: ST
1.a; 2.a; 3.a; 4.a; 5.b; 6.a;
8. translation ceased publication as a separate journal; combined with "Elektrosvyaz'" (in translation) as "Telecommunications and Radio Engineering" beginning 1963, v. 17+; the new combined translation carries different Russian volume numbers for each part, e.g., Part 1 – Telecommunications, v. 25, Part 2 – Radio Engineering, v. 26.

193 RADIOTEKHNIKA I ELEKTRONIKA
(Radio Engineering and Electronics Physics)
Trans. begins with 1957, v. 2+
Monthly
Order from: IEEE or ST
1.a; 2.a; 3.a; 4.a; 5.b; 6.b;
7. 7 mos.;
8. vols. 2–6, no. 7, were translated as "Radio Engineering and Electronics" by PP, and had volume indices.

194 RAZVEDKA I OKHRANA NEDR
(Prospection et Protection du Sous-Sol)
Trans. covers 1959–1962.
Monthly
Order from: BRGGM
7. translation ceased publication.
9.

195 REAKTSIONNAYA SPOSOBNOST' ORGANICHESKIKH SOEDINENII
(Organic Reactivity)
Trans. covers 1966–1970, vols. 3–7.
Quarterly
Order from: PLPC
1.a; 2.a; 3.a; 4.a; 5.a; 6.a;
7. translation ceased publication.

196 REFERATIVNYI ZHURNAL KIBERNETIKA
(Cybernetics Abstracts)
Trans. begins with 1964+
Monthly
Order from: SIC
1.b; 2.c; 3.a; 4.a; 5.b; 6.a;
7. 6 mos.;
8. carries original abstract numbers and translates all abstracts other than those of articles originally published in English, French, or German; the 1964 Russian journal title was "Referativnyi Zhurnal. Matematika, Teoreticheskaya Kibernetika", and the translation was titled "Theoretical Cybernetics Abstracts".

197 REFERATIVNYI ZHURNAL. KORROZIYA I ZASHCHITA OT KORROZII
(Corrosion Control Abstracts)
Trans. begins with 1966, no. 1+
Bimonthly 1966–1969
Monthly 1970+
Order from: SIC
9.

REFERATIVNYI ZHURNAL. MATEMATIKA, TEORETICHESKAYA KIBERNETIKA see "Referativnyi Zhurnal. Kibernetika"

198 REFERATIVNYI ZHURNAL. MEKHANIKA
(Soviet Abstracts: Mechanics)
Trans. covers 1956–1968.
Monthly
Order from: MTRC-OP
9.

EXPLANATION FOR PART I ENTRIES

**   *Indicates that the title has been announced for publication in translation, but has not been so published as of the date of this study (June 1, 1972). Consult publisher for further information.*

Trans. begins — refers to year and volume of the original publication. Unless otherwise indicated, translated volume commences with issue number 1.

Trans. covers — refers to volume (or volumes) translated before translation ceased publication.

OP — following publisher code indicates back issues no longer available from the publisher.

<u>*Explanation of Coded data:*</u>

1. a — Cover-to-cover translation
   b — Selected articles translated

2. a — Full translations of all articles
   b — Some articles fully translated, others abridged or abstracted
   c — Selected articles abridged or abstracted
   d — Articles translated are fully translated, others merely cited or listed

3. a — Translation carries original language numbers
   b — Translation carries other numbers
   c — Translation carries no numbers

4. a — Translation carries original language dates
   b — Translation carries other dates
   c — Translation carries no dates

5. a — Translation identifies original pagination
   b — Translation does not identify original pagination

6. a — Translation has volume index or collective table of contents
   b — Translation has no volume index or collective table of contents

7. "Lag time" between month of issue of original publication and month of issue of translated counter-part (there is a possible variation of one month because of mailing practices both of the original and the translation)

8. Other information (specified)

9. Translation exists but has not been located for verification and evaluation

199 REFERATIVNYI ZHURNAL. METALLURGIYA
(The Abstracts Journal of Metallurgy)
Issued in two parts:
Part A. The Science of Metals
Trans. covers 1957–1963.
Bimonthly
Part B. The Technology of Metals
Trans. covers 1958–1963.
Monthly
Order from: PP
1.b; 2.c; 3.b; 4.a; 5.b; 6.b;
7. translations ceased publication:
8. the PP translated abstracts are provided to and reported in the "ASM Review of Metal Literature" commencing August 1965 through 1967, then in "Metals Abstracts" commencing 1968+.

200 ROCZNIKI CHEMII
(Annals of Chemistry)
Trans. covers 1959–1960, vols. 33–34.
Order from: NTIS
7. translation ceased publication:
9.

201 RUDARSKO-METALURSKI ZBORNIK
(Mining and Metallurgy)
Trans. begins with 1962+
Quarterly
Order from: NTIS
1.a; 2.a; 3.a; 4.a; 5.a; 6. ;
7. 15 mos.

SHU HSUEH HSUEH PAO see
"Acta Mathematica Sinica"

202 SAOPSTENJA INST. ZA VODOPRIVREDU "J. CERNI"
(Works of the Cerni Institute for the Development of Water Resources)
Trans. begins with 1962, v. 9, no. 22+
Quarterly
Order from: NTIS
9.

203 SAVREMENA POLJOPRIVREDA
(Contemporary Agriculture)
Trans. begins with 1966, v. 14+
Monthly
Order from: NTIS
9.

204 SIBIRSKII MATEMATICHESKII ZHURNAL
(Siberian Mathematical Journal)
Trans. begins with 1966, v. 7+
Bimonthly
Order from: PLPC
1.a; 2.a; 3.a; 4.a; 5.a; 6.a;
7. 6 mos.

205 SINTEZY GETEROTSIKLICHESKIKH SOEDINENII
(Synthesis of Heterocyclic Compounds)
Trans. covers 1956–1959, vols. 1–4.
Order from: PLPC
7. original and translation ceased publication;
8. translated issues are only available in joint volumes: 1–2, 3–4;
9.

206 SINTEZY ORGANICHESKIKH SOEDINENII
(Synthesen Organischer Verbindungen)
Trans. begins with 1957, v. 1+
Order from: PV
9.

207 SOVETSKAYA ANTARTICHESKAYA EKSPEDITSIIA, INFORMATSIONNYI BYULLETEN'
(Information Bulletin of the Soviet Antarctic Expedition)
Trans. covers 1958–1961, nos. 1–30.
Order from: AGU
1.a; 2.a; 3.a; 4.a; 5.a; 6.a;
7. original and translation changed title commencing 1961, no. 31;
8. see "Informatsionnyi Byulleten' Sovetskoi Antarticheskoi Ekspeditsii" for successor publication.

208 SOVETSKAYA GEOLOGIYA
(Soviet Geology)
Trans. begins with 1960+
Order from: AGI
8. while the entire Soviet journal is being translated, only selected papers are published in "International Geology Review", but papers not published in "IGR" are available in translation from AGI at a nominal charge.

209 SOVETSKOE ZDRAVOOKHRANENIE
(Soviet Public Health)
Trans. begins with 1966, v. 25+
Monthly
Order from: NTIS
[continued next page]

EXPLANATION FOR PART I ENTRIES

** *Indicates that the title has been announced for publication in translation, but has not been so published as of the date of this study (June 1, 1972). Consult publisher for further information.*

Trans. begins — refers to year and volume of the original publication. Unless otherwise indicated, translated volume commences with issue number 1.

Trans. covers — refers to volume (or volumes) translated before translation ceased publication.

OP — following publisher code indicates back issues no longer available from the publisher.

<u>*Explanation of Coded data:*</u>

1. a — *Cover-to-cover translation*
   b — *Selected articles translated*

2. a — *Full translations of all articles*
   b — *Some articles fully translated, others abridged or abstracted*
   c — *Selected articles abridged or abstracted*
   d — *Articles translated are fully translated, others merely cited or listed*

3. a — *Translation carries original language numbers*
   b — *Translation carries other numbers*
   c — *Translation carries no numbers*

4. a — *Translation carries original language dates*
   b — *Translation carries other dates*
   c — *Translation carries no dates*

5. a — *Translation identifies original pagination*
   b — *Translation does not identify original pagination*

6. a — *Translation has volume index or collective table of contents*
   b — *Translation has no volume index or collective table of contents*

7. *"Lag time" between month of issue of original publication and month of issue of translated counter-part (there is a possible variation of one month because of mailing practices both of the original and the translation)*

8. *Other information (specified)*

9. *Translation exists but has not been located for verification and evaluation*

8. v. 12, 1953, translated as "Sowjetisches Gesundheitwesen", vols. 13–14, 1954–1955, translated as "Sowjetisches Gesundheitsschutz", both published in Berlin.
9.

210 SOVIET OCEANOGRAPHY
[Includes translations of "Doklady Akademii Nauk SSSR" (Oceanology sections) and "Akademii Nauk SSSR. Trudy Morskogo Gidrofizicheskogo Instituta" (Transactions of the Marine Hydrophysical Institute, Academy of Sciences of the USSR)]
Trans. covers 1961–1964.
(Doklady sections 1961–1964,
Morskoi sections 1962–1964).
Quarterly
Order from: AGU
1.a; 2.a; 3.a; 4.a; 5.a; 6.b;
7. translation ceased publication.

211 STAL'
(Stal' in English)
Trans. covers 1959–1970, vols. 17–30.
Monthly
Order from: BISI
1.a; 2.a; 3.a; 4.a; 5.a; 6.a;
7. succeeded by "Steel in the USSR", see Part II;
8. another translation, in German, "Stahl", is available from DBEI beginning 1961, v. 21+.

212 STANKI I INSTRUMENT
(Machines and Tooling)
Trans. begins with 1959, v. 30+
Monthly
Order from: PERA
1.a; 2.a; 3.a; 4.a; 5.a; 6. ;
7. 4 mos.

213 STAUB-REINHALTUNG DER LUFT
[in English]
Trans. begins with 1965, v. 25+
Monthly
Order from: NTIS
1.a; 2.a; 3.a; 4.a; 5.b; 6. ;
7. 19 mos.

214 STEKLO I KERAMIKA
(Glass and Ceramics)
Trans. begins with 1956, v. 13+
Monthly
Order from: PLPC
1.a; 2.a; 3.a; 4.a; 5.a; 6.a;
7. 6 mos.

215 SVAROCHNOE PROIZVODSTVO
(Welding Production)
Trans. begins with 1959, v. 5, no. 4+
Monthly
Order form: BWRA
9.

216 TECHNISCHE MITTEILUNGEN WERKSBERICHTE
(Krupp Technical Review)
Trans. begins with 1964, v. 22+
Order from: KRUPP
8. issued in English simultaneously with original German edition.

217 TEORETICHESKAYA I EKSPERIMENTAL'NAYA KHIMIYA
(Theoretical and Experimental Chemistry)
Trans. begins with 1965, v. 1+
Bimonthly
Order from: PLPC
1.a; 2.a; 3.a; 4.a; 5.a; 6.a;
7. the present publisher of this translation is making every effort to reduce previously existing translation delays and it is deemed advisable not to cite any specific "lag time" in this edition of the GUIDE.

218 TEORETICHESKAYA I MATEMATICHESKAYA FIZIKI
(Theoretical and Mathematical Physics)
Trans. begins with 1969, v. 1+
Monthly
Order from: PLPC
1.a; 2.a; 3.a; 4.a; 5.a; 6.a;
7. 6 mos.

219 TEORETICHESKIE OSNOVY KHIMICHESKOI TEKHNOLOGII
(Theoretical Foundations of Chemical Engineering)
Trans. begins with 1967, v. 1+
Bimonthly
Order from: PLPC
1.a; 2.a; 3.a; 4.a; 5.a; 6. ;
7. 6 mos.

220 TEORIYA VEROYATNOSTEI I EE PRIMENENIYA
(Theory of Probability and its Applications)
Trans. begins with 1956, v. 1+
Quarterly
Order from: SIAM
1.a; 2.a; 3.a; 4.a; 5.b; 6.a;
7. 6 mos.

EXPLANATION FOR PART I ENTRIES

** *Indicates that the title has been announced for publication in translation, but has not been so published as of the date of this study (June 1, 1972). Consult publisher for further information.*

Trans. begins — refers to year and volume of the original publication. Unless otherwise indicated, translated volume commences with issue number 1.

Trans. covers — refers to volume (or volumes) translated before translation ceased publication.

OP — following publisher code indicates back issues no longer available from the publisher.

<u>Explanation of Coded data:</u>

1. a — Cover-to-cover translation
   b — Selected articles translated

2. a — Full translations of all articles
   b — Some articles fully translated, others abridged or abstracted
   c — Selected articles abridged or abstracted
   d — Articles translated are fully translated, others merely cited or listed

3. a — Translation carries original language numbers
   b — Translation carries other numbers
   c — Translation carries no numbers

4. a — Translation carries original language dates
   b — Translation carries other dates
   c — Translation carries no dates

5. a — Translation identifies original pagination
   b — Translation does not identify original pagination

6. a — Translation has volume index or collective table of contents
   b — Translation has no volume index or collective table of contents

7. "Lag time" between month of issue of original publication and month of issue of translated counter-part (there is a possible variation of one month because of mailing practices both of the original and the translation)

8. Other information (specified)

9. Translation exists but has not been located for verification and evaluation

221 TEPLOENERGETIKA
(Thermal Engineering)
Trans. begins with 1962, v. 9+
Monthly
Order from: PP
1.a; 2.a; 3.a; 4.a; 5.a; 6. ;
7. 8 mos.;
8. vols. 9–11, 1962–1964, are available, abstracts only, from FP under the title "Heat and Power".

222 TEPLOFIZIKA VYSOKIKH TEMPERATUR
(High Temperature)
Trans. begins with 1963, v. 1+
Bimonthly
Order from: AIP
1.a; 2.a; 3.a; 4.a; 5.a; 6.a;
7. 6 mos.

223 TRENIE I IZNOS V MASHINAKH
(Friction and Wear in Machinery)
Trans. covers 1956–1967, vols. 11–19.
Irregular
Order from: ASME
8. Russian original carries no volume numbers;
9.

224 TRUDY FIZICHESKOGO INSTITUTA IMENI P. N. LEBEDEVA, AKADEMII NAUK SSSR
(The Lebedev Physics Institute Series)
Trans. begins with 1963, v. 25+
Irregular
Order from: PLPC
1.a; 2.a; 3.a; 4.a; 5.a; 6.a;
7. 12 mos. average;
8. individual volumes carry individual titles and are issued as separate volumes.

225 TRUDY GEOFIZICHESKOGO INSTITUTA AKADEMII NAUK SSSR
(Soviet Research in Geophysics)
Trans. covers 1957, nos. 37–40.
Order from: PLPC
7. translation ceased publication;
9.

226 TRUDY INSTITUTA VYSSHEI NERVNOI DEYATEL'NOSTI. SERIYA FIZIOLOGICHESKAYA
(Works of the Institute of Higher Nervous Activity. Physiological Series)
Trans. covers 1955–1962, vols. 1–7.
Order form: NTIS
[continued next column]

7. translation ceased publication;
9.

227 TRUDY INSTITUTA VYSSHEI NERVNOI DEYATEL'NOSTI. SERIYA PATOFIZIOLOGICHESKAYA
(Works of the Institute of Higher Nervous Activity. Pathophysiological Series)
Trans. covers 1955–1962, vols. 1–10.
Order from: NTIS
1.a; 2.a; 3.a; 4.a; 5.b; 6.b;
7. translation ceased publication.

228 TRUDY MATEMATICHESKOGO INSTITUTA IMENI V. A. STEKLOVA
(Proceedings of the Steklov Institute of Mathematics of the Academy of Sciences of the USSR)
Trans. begins with 1966, v. 74+
Irregular
Order from: AMS
9.

229 TRUDY MOSKOVSKOGO MATEMATICHESKOGO OBSHCHESTVA
(Transactions of the Moscow Mathematical Society)
Trans. begins with 1963, v. 12+
Annual
Order from: AMS
7. 12 mos.

TRUDY MORSKOGO GIDROFIZICHESKOGO INSTITUTA. AKADEMII NAUK SSSR see "Soviet Oceanography"

230 TSEMENT
(Cement)
Trans. covers 1956–1957, vols. 22–23, no. 2.
Bimonthly
Order from: PLPC-OP
7. translation ceased publication;
9.

231 TSITOLOGIYA
(Cytology)
Trans. covers 1959–1961, vols. 1–3.
Bimonthly
Order from: NTIS
7. translation ceased publication;
9.

EXPLANATION FOR PART I ENTRIES

** Indicates that the title has been announced for publication in translation, but has not been so published as of the date of this study (June 1, 1972). Consult publisher for further information.

Trans. begins — refers to year and volume of the original publication. Unless otherwise indicated, translated volume commences with issue number 1.

Trans. covers — refers to volume (or volumes) translated before translation ceased publication.

OP — following publisher code indicates back issues no longer available from the publisher.

Explanation of Coded data:

1. a — *Cover-to-cover translation*
   b — *Selected articles translated*

2. a — *Full translations of all articles*
   b — *Some articles fully translated, others abridged or abstracted*
   c — *Selected articles abridged or abstracted*
   d — *Articles translated are fully translated, others merely cited or listed*

3. a — *Translation carries original language numbers*
   b — *Translation carries other numbers*
   c — *Translation carries no numbers*

4. a — *Translation carries original language dates*
   b — *Translation carries other dates*
   c — *Translation carries no dates*

5. a — *Translation identifies original pagination*
   b — *Translation does not identify original pagination*

6. a — *Translation has volume index or collective table of contents*
   b — *Translation has no volume index or collective table of contents*

7. *"Lag time" between month of issue of original publication and month of issue of translated counter-part (there is a possible variation of one month because of mailing practices both of the original and the translation)*

8. *Other information (specified)*

9. *Translation exists but has not been located for verification and evaluation*

232 TSVETNYE METALLY
(The Soviet Journal of Nonferrous Metals)
Trans. begins with 1960, v. 33+
Monthly
Order from: PS
1.a; 2.a; 3.a & b; 4.a & b; 5.a; 6.a;
7. 28 mos.;
8. Soviet journal carries no volume numbers but translation carries a separate Soviet volume number and a separate translation volume number.

233 UKRAINSKII FIZICHESKII ZHURNAL
(Ukrainian Physics Journal)
Trans. covers 1967–1968, vols. 12–13.
Order v. 12 from NTIS, v. 13 from AIP

UKRAINSKII FIZICHNII ZHURNAL see "Ukrainskii Fizicheskii Zhurnal"

234 UKRAINSKII KHIMICHESKII ZHURNAL
(Soviet Progress in Chemistry)
Trans. begins with 1966, v. 32+
Monthly
Order from: APr
1.a; 2.a; 3.a; 4.a; 5.a; 6.a;
7. the present publisher of this translation is making every effort to reduce previously existing translation delays and it is deemed advisable not to cite any specific "lag time" in this edition of the GUIDE.
8. v. 28, nos. 1–9, 1962, was translated and is available from NTIS under the title "Ukrainian Journal of Chemistry".

235 UKRAINSKII MATEMATICHESKII ZHURNAL
(Ukrainian Mathematical Journal)
Trans. begins with 1967, v. 19+
Bimonthly
Order from: PLPC
1.a; 2.a; 3.a; 4.a; 5.a; 6.a;
7. 8 mos.

236 USPEKHI FIZICHESKIKH NAUK
(Soviet Physics–Uspekhi)
Trans. begins with 1957, v. 61+
Bimonthly
Order from: AIP
1.a; 2.a; 3.b; 4.a; 5.a; 6.a;
7. 6 mos.;
8. vols. 61–65, 1957–1958, were translated as "Advances in the Physical Sciences", and issued through NTIS.

237 USPEKHI FIZIOLOGICHESKIKH NAUK
(Progress in Physiological Sciences)
Trans. begins with 1970, v. 1+
Quarterly
Order from: PLPC
1.a; 2.a; 3.a; 4.a; 5.a; 6.a;
7. 6 mos.

238 USPEKHI KHIMII
(Russian Chemical Reviews)
Trans. begins with 1960, v. 24+
Monthly
Order from: CS
1.a; 2.a; 3.a; 4.a; 5.a; 6.a;
7. 9 mos.

239 USPEKHI MATEMATICHESKIKH NAUK
(Russian Mathematical Surveys)
Trans. begins with 1960, v. 15+
Bimonthly
Order from: MACM
1.b; 2.d; 3.a; 4.a; 5.b; 6.a;
7. 12 mos.

240 USPEKHI SOVREMENNOI BIOLOGII
(Russian Review of Biology)
Trans. covers 1959–1960, vols. 48–50, no. 3.
Bimonthly
Order from: NLL
7. translation ceased publication:
9.

241 VESNIK ZAVODA ZA GEOLOSKA I GEOFIZICHESKA ISTRAZHIVANIJA
(Bulletin of the Institute for Geological and Geophysical Research)
Trans. begins with 1962+
Order from: NTIS
7. 12 mos.;
8. an annual issued in three separate series: (A)–Geology, (B)–Engineering Geology, (C)–Applied Geophysics.

242 VESTNIK AKADEMII MEDITSINSKIKH NAUK SSSR
(Vestnik of the USSR Academy of Medical Sciences)
Trans. covers 1962–1964, vols. 17–19.
Monthly
Order from: NTIS
[continued next page]

EXPLANATION FOR PART I ENTRIES

** *Indicates that the title has been announced for publication in translation, but has not been so published as of the date of this study (June 1, 1972). Consult publisher for further information.*

Trans. begins — refers to year and volume of the original publication. Unless otherwise indicated, translated volume commences with issue number 1.

Trans. covers — refers to volume (or volumes) translated before translation ceased publication.

OP — following publisher code indicates back issues no longer available from the publisher.

Explanation of Coded data:

1. a — Cover-to-cover translation
   b — Selected articles translated

2. a — Full translations of all articles
   b — Some articles fully translated, others abridged or abstracted
   c — Selected articles abridged or abstracted
   d — Articles translated are fully translated, others merely cited or listed

3. a — Translation carries original language numbers
   b — Translation carries other numbers
   c — Translation carries no numbers

4. a — Translation carries original language dates
   b — Translation carries other dates
   c — Translation carries no dates

5. a — Translation identifies original pagination
   b — Translation does not identify original pagination

6. a — Translation has volume index or collective table of contents
   b — Translation has no volume index or collective table of contents

7. "Lag time" between month of issue of original publication and month of issue of translated counter-part (there is a possible variation of one month because of mailing practices both of the original and the translation)

8. Other information (specified)

9. Translation exists but has not been located for verification and evaluation

8. vols. 17–18, 1962–1963, were translated as "Herald of the Academy of Medical Sciences", and the following issues are available: v. 17, nos. 2, 4–6; v. 18, nos. 1–8, 10–12;
9.

243 VESTNIK MASHINOSTROENIYA
(Russian Engineering Journal)
Trans. begins with 1959, v. 39, no. 4+
Monthly
Order from: PERA
1.a; 2.a; 3.a; 4.a; 5.a; 6.a;
7. 3 mos.;
8. each index is issued as a separate item.

244 VESTNIK MOSKOVSKOGO UNIVERSITETA. SERIYA FIZIKA, ASTRONOMIYA
(Moscow University Physics Bulletin)
Trans. begins with 1966+
Bimonthly
Order from: APr
1.a; 2.a; 3.a; 4.a; 5.a; 6.a;
7. the present publisher of this translation is making every effort to reduce previously existing translation delays and it is deemed advisable not to cite any specific "lag time" in this edition of the GUIDE.

245 VESTNIK MOSKOVSKOGO UNIVERSITETA. SERIYA KHIMIYA
(Moscow University Chemistry Bulletin)
Trans. begins with 1966, v. 21+
Bimonthly
Order form: APr
1.a; 2.a; 3.a; 4.a; 5.a; 6.a;
7. the present publisher of this translation is making every effort to reduce previously existing translation delays and it is deemed advisable not to cite any specific "lag time" in this edition of the GUIDE.

246 VESTNIK MOSKOVSKOGO UNIVERSITETA. PERVAIYA SERIYA – MATEMATIKA, MEKHANIKA
(Moscow University Mathematics Bulletin)
(Moscow University Mechanics Bulletin)
Trans. begins with 1966, v. 21+
Quarterly
Order from: APr
1.a; 2.a; 3.a; 4.a; 5.a; 6.a;
[continued next column]

7. the present publisher of this translation is making every effort to reduce previously existing translation delays and it is deemed advisable not to cite any specific "lag time" in this edition of the GUIDE.
8. original is not two separate publications, though each subject area is a separate section, but the translator is issuing the sections as separate publications.

247 VESTNIK SVYAZI
(Herald of Communications)
Trans. covers 1954, v. 14.
Monthly
Order from: NTIS
8. the following issues are also available in translation; 1947, v. 7, nos. 1–12; 1948, v. 8, nos. 1–11; 1952, v. 12, nos. 1, 6–8, 10; 1953, v. 13, nos. 6–12;
9.

248 VODNYE RESURSY**
(Water Resources)
Trans. to begin with Russian 1972 issue no. 1
Order from: PLPC

249 VOENNO-MEDITSINSKII ZHURNAL
(Military Medical Journal)
Trans. covers 1958, no. 10– 1961, no. 12.
Monthly
Order from: NTIS
7. translation ceased publication:
9.

250 VOPROSY IKHTIOLOGII
(Journal of Icthiology)
Trans. begins with 1968, v. 8+
Bimonthly
Order from: AFS
9.

251 VOPROSY KOSMOGONII
(Problems of Cosmogony)
Trans. covers 1952–1962, vols. 1–8.
Order from: NTIS
7. translation ceased publication;
9.

252 VOPROSY MIKROPALEONTOLOGII
(Questions de Micropaleontologie. Paris)
Trans. begins with 1956+
Order from: BRGGM
9.

EXPLANATION FOR PART I ENTRIES

** *Indicates that the title has been announced for publication in translation, but has not been so published as of the date of this study (June 1, 1972). Consult publisher for further information.*

Trans. begins — refers to year and volume of the original publication. Unless otherwise indicated, translated volume commences with issue number 1.

Trans. covers — refers to volume (or volumes) translated before translation ceased publication.

OP — following publisher code indicates back issues no longer available from the publisher.

Explanation of Coded data:

1. *a — Cover-to-cover translation*
   b — Selected articles translated

2. *a — Full translations of all articles*
   b — Some articles fully translated, others abridged or abstracted
   c — Selected articles abridged or abstracted
   d — Articles translated are fully translated, others merely cited or listed

3. *a — Translation carries original language numbers*
   b — Translation carries other numbers
   c — Translation carries no numbers

4. *a — Translation carries original language dates*
   b — Translation carries other dates
   c — Translation carries no dates

5. *a — Translation identifies original pagination*
   b — Translation does not identify original pagination

6. *a — Translation has volume index or collective table of contents*
   b — Translation has no volume index or collective table of contents

7. *"Lag time" between month of issue of original publication and month of issue of translated counter-part (there is a possible variation of one month because of mailing practices both of the original and the translation)*

8. *Other information (specified)*

9. *Translation exists but has not been located for verification and evaluation*

253 VOPROSY ONKOLOGII
(Problems of Oncology)
Trans. covers 1957–1961, vols. 3–7.
Order from: PP
7. translation ceased publication:
9.

254 VOPROSY PSIKHOLOGII
(Problems of Psychology)
Trans. covers 1960, v. 6, nos. 1–4.
Quarterly
Order from: PP
7. translation ceased publication.

255 VOPROSY VIRUSOLOGII
(Problems of Virology)
Trans. covers 1957–1961, vols. 2–6.
Bimonthly
Order from: PP
1.a; 2.a; 3.a; 4.a; 5.a; 6.a;
7. translation ceased publication.

256 VYCHISLITEL'NYE SISTEMY.
SBORNIK TRUDOV
(Computer Elements and Systems)
Trans. covers 1962–1963, nos. 1–9.
Order from: NTIS
9.

257 VYSOKOMOLEKULYARNYE SOEDINENIYA
(Polymer Science USSR)
Trans. begins with 1959, v. 1+
Monthly
Order from: PP
1.b; 2.a; 3.a; 4.a; 5.a; 6.a;
7. 10 mos.;
8. only selected articles were translated for 1959–1962.

258 WERKSTATT UND BETRIEB
(Machine Tool Engineering and Production News)
English translation of technical articles published within the enlarged volume but on different colored pages (the standard German edition does not have these pages). Publication begins with 1960+
Order from: WERK

WU LI HSUEH PAO see
"Acta Physica Sinica"

259 YADERNAYA FIZIKA
(Soviet Journal of Nuclear Physics)
Trans. begins with 1965, v. 1+
Monthly
Order from: AIP
1.a; 2.a; 3.a; 4.a; 5.a; 6.a;
7. 6 mos.

260 ZASHCHITA METALLOV
(Protection of Metals)
Trans. begins with 1965, v. 1+
Bimonthly
Order from: PLPC
1.a; 2.a; 3.a; 4.a; 5.a; 6.a;
7. 6 mos.

261 ZAVODSKAYA LABORATORIYA
(Industrial Laboratory)
Trans. begins with 1958, v. 24+
Monthly
Order from: PLPC
1.a; 2.a; 3.a; 4.a; 5.a; 6.a;
7. 5 mos.

262 ZhETF PIS'MA V REDAKTSIYU
(JETP Letters)
Trans. begins with 1965, v. 1+
Semimonthly
Order from: AIP
1.a; 2.a; 3.a; 4.a; 5.a; 6.a;
7. 6 weeks

263 ZHURNAL ANALITICHESKOI KHIMII
(Journal of Analytical Chemistry of the USSR)
Trans. begins with 1952, v. 7+
Monthly
Order from: PLPC
1.a; 2.a; 3.a; 4.a; 5.a; 6.a;
7. 6 mos.

264 ZHURNAL EKSPERIMENTAL'NOI I
TEORETICHESKOI FIZIKI
(Soviet Physics–JETP)
Trans. begins with 1955, v. 28+
Monthly
Order from: AIP
1.a; 2.a; 3.a; 4.a; 5.a; 6.a;
7. 6 mos.

EXPLANATION FOR PART I ENTRIES

** Indicates that the title has been announced for publication in translation, but has not been so published as of the date of this study (June 1, 1972). Consult publisher for further information.

Trans. begins — refers to year and volume of the original publication. Unless otherwise indicated, translated volume commences with issue number 1.

Trans. covers — refers to volume (or volumes) translated before translation ceased publication.

OP — following publisher code indicates back issues no longer available from the publisher.

Explanation of Coded data:

1. a — *Cover-to-cover translation*
   b — *Selected articles translated*

2. a — *Full translations of all articles*
   b — *Some articles fully translated, others abridged or abstracted*
   c — *Selected articles abridged or abstracted*
   d — *Articles translated are fully translated, others merely cited or listed*

3. a — *Translation carries original language numbers*
   b — *Translation carries other numbers*
   c — *Translation carries no numbers*

4. a — *Translation carries original language dates*
   b — *Translation carries other dates*
   c — *Translation carries no dates*

5. a — *Translation identifies original pagination*
   b — *Translation does not identify original pagination*

6. a — *Translation has volume index or collective table of contents*
   b — *Translation has no volume index or collective table of contents*

7. *"Lag time" between month of issue of original publication and month of issue of translated counter-part (there is a possible variation of one month because of mailing practices both of the original and the translation)*

8. *Other information (specified)*

9. *Translation exists but has not been located for verification and evaluation*

265 ZHURNAL EVOLYUTSIONNOI BIOKHIMI I FIZIOLOGI
(Journal of Evolutionary Biochemistry and Physiology)
Trans. begins with 1969, v. 5+
Bimonthly
Order from: PLPC
1.a; 2.a; 3.a; 4.a; 5.a; 6.a;
7. 6 mos.

266 ZHURNAL FIZICHESKOI KHIMII
(Russian Journal of Physical Chemistry)
Trans. begins with 1959, v. 33+
Monthly
Order from: CS
1.a; 2.a; 3.a; 4.a; 5.a; 6.a;
7. 6 mos.;
8. v. 33, nos. 1–6 are only available as abstracts.

267 ZHURNAL MIKROBIOLOGII, EPIDEMIOLOGII I IMMUNOBIOLOGII
(Journal of Microbiology, Epidemiology and Immunobiology)
Trans. covers 1957–1961, vols. 28–32.
Order from: PP
1.a; 2.a; 3.a; 4.a; 5.a; 6.a;
7. translation ceased publication.

268 ZHURNAL NEORGANICHESKOI KHIMII
(Russian Journal of Inorganic Chemistry)
Trans. begins with 1956, v. 1+
Monthly
Order from: CS
1.a; 2.a; 3.a; 4.a; 5.a; 6.a;
7. 6 mos.;
8. vols. 1–3, 1956–1958, titled "Journal of Inorganic Chemistry", are available from NTIS.

269 ZHURNAL OBSHCHEI KHIMII
(Journal of General Chemistry of the USSR)
Trans. begins with 1949, v. 19+
Monthly
Order from: PLPC
1.a; 2.a; 3.a; 4.a; 5.a; 6.a;
7. 6 mos.

270 ZHURNAL ORGANICHESKOI KHIMII
(Journal of Organic Chemistry of the USSR)
Trans. begins with 1965, v. 1+
Monthly
Order from: PLPC
1.a; 2.a; 3.a; 4.a; 5.a; 6.a;
7. 6 mos.

271 ZHURNAL PRIKLADNOI KHIMII
(Journal of Applied Chemistry of the USSR)
Trans. begins with 1950, v. 23+
Monthly
Order from: PLPC
1.a; 2.a; 3.a; 4.a; 5.a; 6.a;
7. 6 mos.

272 ZHURNAL PRIKLADNOI MEKHANIKI I TEKHNICHESKOI FIZIKI
(Journal of Applied Mechanics and Technical Physics)
Trans. begins with 1965, v. 6+
Bimonthly
Order from: PLPC
1.a; 2.a; 3.a; 4.a; 5.a; 6.a;
7. the present publisher of this translation is making every effort to reduce previously existing translation delays and it is deemed advisable not to cite any specific "lag time" in this edition of the GUIDE.

273 ZHURNAL PRIKLADNOI SPEKTROSKOPII
(Journal of Applied Spectroscopy)
Trans. begins with 1965, v. 2+
Monthly
Order from: PLPC
1.a; 2.a; 3.a; 4.a; 5.a; 6.a;
7. the present publisher of this translation is making every effort to reduce previously existing translation delays and it is deemed advisable not to cite any specific "lag time" in this edition of the GUIDE.

274 ZHURNAL STRUKTURNOI KHIMII
(Journal of Structural Chemistry)
Trans. begins with 1960, v. 1+
Bimonthly
Order from: PLPC
1.a; 2.a; 3.a; 4.a; 5.a; 6.a;
7. 6 mos.

275 ZHURNAL TEKHNICHESKOI FIZIKI
(Soviet Physics–Technical Physics)
Trans. begins with 1956, v. 26+
Monthly
Order from: AIP
1.a; 2.a; 3.a; 4.a; 5.a; 6.a;
7. 6 mos.

EXPLANATION FOR PART I ENTRIES

** Indicates that the title has been announced for publication in translation, but has not been so published as of the date of this study (June 1, 1972). Consult publisher for further information.

Trans. begins — refers to year and volume of the original publication. Unless otherwise indicated, translated volume commences with issue number 1.

Trans. covers — refers to volume (or volumes) translated before translation ceased publication.

OP — following publisher code indicates back issues no longer available from the publisher.

Explanation of Coded data:

1. a — Cover-to-cover translation
   b — Selected articles translated

2. a — Full translations of all articles
   b — Some articles fully translated, others abridged or abstracted
   c — Selected articles abridged or abstracted
   d — Articles translated are fully translated, others merely cited or listed

3. a — Translation carries original language numbers
   b — Translation carries other numbers
   c — Translation carries no numbers

4. a — Translation carries original language dates
   b — Translation carries other dates
   c — Translation carries no dates

5. a — Translation identifies original pagination
   b — Translation does not identify original pagination

6. a — Translation has volume index or collective table of contents
   b — Translation has no volume index or collective table of contents

7. "Lag time" between month of issue of original publication and month of issue of translated counter-part (there is a possible variation of one month because of mailing practices both of the original and the translation)

8. Other information (specified)

9. Translation exists but has not been located for verification and evaluation

276 ZHURNAL VSESOYUZNOGO KHIMICHESKOGO OBSHCHESTVA IMENI D. I. MENDELEEVA
(Mendeleev Chemistry Journal)
Trans. begins with 1966, v. 11+
Bimonthly
Order from: APr
1.a; 2.a; 3.a; 4.a; 5.a; 6.a;
7. the present publisher of this translation is making every effort to reduce previously existing translation delays and it is deemed advisable not to cite any specific "lag time" in this edition of the GUIDE.

277 ZHURNAL VYCHISLITEL'NOI MATEMATIKI I MATEMATICHESKOI FIZIKI
(USSR Computational Mathematics and Mathematical Physics)
Trans. begins with 1961, v. 1+
Quarterly 1961,
Bimonthly 1962+
Order from: PP
1.b; 2.a; 3.b; 4.a; 5.a; 6. ;
7. 29 mos.;
8. selected articles are carried in "International Journal of Computer Mechanics".

278 ZHURNAL VYSSHEI NERVNOI DEYATEL'NOSTI IMENI I. P. PAVLOVA
(Pavlov Journal of Higher Nervous Activity)
Trans. covers 1952–1961, vols. 2–11.
Bimonthly
Order from: PP and EL
1.a; 2.a; 3.a; 4.a; 5.b; 6.a;
7. translation ceased publication;
8. 1952–1961, vols. 2–11, translated as "Pawlow-Zeitschrift fur hohere Nerventatigkeit" by VVG; 1958–1961, vols. 8–11, translated by PP under English title; 1961, v. 11, translated by ST under same title.

PART II

SELECTIONS, COLLECTIONS, AND OTHER "TRANSLATION" JOURNALS

EXPLANATION FOR PART II ENTRIES

Most journals here-in listed are not cover-to-cover translations of specific original language journals.

(*) Indicates publications which gather their material from a variety of sources and, while useful as an 'alerting service', cannot be relied upon as a retrospective retrieval source.

Some publications here-in listed are not translations as the translated edition is issued simultaneously with the original, or carries the translated section within the original.

Some publications are listed to correct the erroneous impression that they are translations when they are not.

A few titles are listed in this section because of the significant incompleteness of the translations.

1. **ABSTRACTS OF BULGARIAN SCIENTIFIC LITERATURE (*)**
   Publication begins with 1958+
   Issued in English in several series.
   Order from: TNTID

2. **AIAA JOURNAL: RUSSIAN SUPPLEMENT (*)**
   Publication begins with 1959+
   Publication was part of ARS Journal from 1959, v. 29, no. 10– 1962, v. 32, no. 12; began with AIAA Journal 1963, v. 1+
   Order from: AIAA

3. **AKADEMIYA MEDITSINSKIKH NAUK SSSR**
   (Medical Publishing Plan, Academy of Medical Sciences, USSR)
   Translation begins with 1965+
   Order from: SIC

4. **AKADEMIYA NAUK SSSR – TEMATICHESKIY PLAN**
   (Publishing Plan, Academy of Sciences, USSR)
   Trans. begins with 1964+
   Order from: SIC

5. **AMERICAN MATHEMATICAL SOCIETY TRANSLATIONS (*)**
   Publication begins with 1949+
   Order from: AMS

6. **AMFETEX JOURNAL OF JAPANESE ELECTRONICS AND POWER (AN ABSTRACTS JOURNAL) (*) ***
   Order from: AMFETEX

7. **ASTRONOMICAL NEWS LETTER (*)**
   Publication covers 1948–1961, nos. 37–109.
   Order from: LAL

8. **AUTOMATION EXPRESS (*)**
   Publication begins with 1958, v. 1+
   Order from: IPI

9. **CZECHOSLOVAK ENGINEERING SCIENCES ABSTRACTS (*)**
   Publication begins with 1970, v. 1+
   Order from: SIC

10. **DOKLADY AKADEMII NAUK SSSR**
    (International Edition)
    Publication covers 1922–1947, Old Series: Series A (Science Sections), 1922–1933,
    [continued next column]
    nos. 1–158, New Series: 1933–1947, vols. 1–56 (excluding v. 50).
    Both series were originally published in western European languages, primarily English, French, and German, not all in any one language and no article in more than one language.
    Order from: KRCL

11. **DRAHT-WELT**
    (Wire World International)
    Publication begins with 1959
    Order from: WWI

12. **EAST EUROPEAN SCIENTIFIC ABSTRACTS (*)**
    Publication begins with 1964+
    Order from: CFSTI

13. **ELECTRONICS EXPRESS (*)**
    Publication begins with 1958+
    Order from: IPI

14. **FEDERATION PROCEEDINGS TRANSLATION SUPPLEMENT (*)**
    Publication of a Supplement as Part II of the Proceedings begins with 1963, v. 22+
    Order from: FASEB

15. **FLUID MECHANICS – SOVIET RESEARCH (*)**
    Publication to begin with Jan. 1972
    Order from: ASME or ST

16. **GEOCHEMISTRY INTERNATIONAL (*)**
    Publication begins with 1964+
    Order from: AGI

17. **HEAT TRANSFER – JAPANESE RESEARCH (*)**
    Publication to begin with Jan. 1972
    Order from: ASME or ST

18. **HEAT TRANSFER – SOVIET RESEARCH (*)**
    Publication to begin with Jan. 1969
    Bimonthly
    Order from: ASME or ST

19. **HUNGARIAN TECHNICAL ABSTRACTS (*)**
    Publication begins with 1949+
    Order from: HTCBN

EXPLANATION FOR PART II ENTRIES

Most journals here-in listed are not cover-to-cover translations of specific original language journals.

(*) Indicates publications which gather their material from a variety of sources and, while useful as an 'alerting service', cannot be relied upon as a retrospective retrieval source.

Some publications here-in listed are not translations as the translated edition is issued simultaneously with the original, or carries the translated section within the original.

Some publications are listed to correct the erroneous impression that they are translations when they are not.

A few titles are listed in this section because of the significant incompleteness of the translations.

20. **INTERNATIONAL CHEMICAL ENGINEERING (*)**
    Publication begins with 1961+
    Order from: ICE

21. **INTERNATIONAL GEOLOGY REVIEW (*)**
    Publication begins with 1959+
    Order from: AGI

    JAPANESE JOURNAL OF APPLIED PHYSICS
    see "Oyo Butsuri"

22. **JOURNAL OF PHYSICS (in English)**
    Publication covers 1939–1947, vols. 1–11.
    Order from: Akademii Nauk SSSR-OP

23. **METEOROLOGIYA I GIDROLOGIYA**
    (Meteorology and Hydrology)
    Trans. begins with 1966, vol. 9
    Bimonthly
    Order from: NTIS

24. **NAUCHNO-TEKHNICHESKAYA INFORMATSIYA**
    (Automatic Documentation and Mathematical Linguistics)
    Selected major articles
    Quarterly
    Order from: APr

25. **NEUROSCIENCE AND BEHAVIORAL PHYSIOLOGY (*)**
    Publication commences August 1971
    Order from: FASEB or ST

26. **NEUROSCIENCES TRANSLATIONS (*)**
    Publication covers 1969–1971, nos. 1–16.
    Quarterly
    Order from: FASEB
    Title changed to "Neuroscience and Behavioral Physiology" in Aug. 1971.

27. **NTZ – NACHRICHTENISCHE ZEITSCHRIFT**
    (NTZ – Communications Journal)
    Trans. begins with 1962+
    Bimonthly
    Order from: NTZ

28. **NUCLEAR SCIENCE ABSTRACTS OF JAPAN [in English] (*)**
    Publication begins with 1961+
    Order from: JAERI

29. **NUCLEAR FUSION**
    Translation of non-English articles of the IAEA Journal.
    Publication begins with 1965, v. 5+
    Order from: CFSTI

30. **OYO BUTSURI**
    (Japanese Journal of Applied Physics [in English])
    This publication has carried only English language articles since its inception and has no relation to any Japanese publication.
    Publication begins with 1965, v. 1+
    Order from: UT

31. **POLISH MEDICAL JOURNAL (*)**
    Publication begins with 1962, v. 1+
    Order from: CFSTI

32. **POLISH TECHNICAL ABSTRACTS (*)**
    (In English)
    Publication begins with 1951+
    Order from: CIINTE

33. **POWER EXPRESS (*)**
    Publication begins with 1961+
    Order from: IPI

34. **REVIEW OF CZECHOSLOVAK MEDICINE (*)**
    Publication begins with 1955+
    Order from: ARTIA

35. **REVIEW OF THE ELECTRICAL COMMUNICATIONS LABORATORY [Japan] (*)**
    Order from: ECL

36. **ROMANIAN SCIENTIFIC ABSTRACTS (*)**
    Publication begins with 1964, v. 1+
    Order from: ARSRCDS

37. **RUSSIAN PHYSICS QUARTERLY (*)**
    Publication covers 1963–1965, vols. 1–2.
    Order from: IPI

38. **SCIENCE PERIODICALS FROM MAINLAND CHINA (*)**
    Publication begins with 1965, v. 1+
    Order from: NFSAIS

EXPLANATION FOR PART II ENTRIES

Most journals here-in listed are not cover-to-cover translations of specific original language journals.

(*) Indicates publications which gather their material from a variety of sources and, while useful as an 'alerting service', cannot be relied upon as a retrospective retrieval source.

Some publications here-in listed are not translations as the translated edition is issued simultaneously with the original, or carries the translated section within the original.

Some publications are listed to correct the erroneous impression that they are translations when they are not.

A few titles are listed in this section because of the significant incompleteness of the translations.

39. **SELECTED TRANSLATIONS IN MATHEMATICAL STATISTICS AND PROBABILITY (*)**
    Publication begins with 1961, v. 1+
    Order from: AMS

40. **SOVIET GEOGRAPHY: REVIEW AND TRANSLATION (*)**
    Publication begins with 1960, v. 1+
    Order from: AGS

41. **SOVIET HYDROLOGY: SELECTED PAPERS (*)**
    Publication begins with 1961+
    Order from: AGU

42. **SOVIET METAL TECHNOLOGY (*)**
    Publication covers 1958–1959, vols. 1–2.
    Order from: PS

43. **SOVIET PSYCHIATRY (*)**
    Publication begins with 1966, v. 5+
    Order from: IASP

44. **SOVIET PSYCHOLOGY (*)**
    Publication begins with 1966, v. 5+
    Order from: IASP

45. **SOVIET PSYCHOLOGY AND PSYCHIATRY (*)**
    Publication covers 1962–1965, vols. 1–4.
    Order from: IASP

46. **SOVIET TECHNOLOGY DIGEST (*)**
    Publication covers 1959–1961, vols. 1–3.
    Order from: PP

47. **STEEL IN THE USSR (*)**
    Publication begins with Jan. 1971
    Monthly
    Selected articles are translated from "Stal'" and "Izv. V.U.Z., Chernaya Metallurgiya"
    Order from: BISI

48. **TECHNICAL PHYSICS OF THE USSR, LENINGRAD (in English)**
    Publication covers 1934–1938, vols. 1–5.
    Out-of-print.

49. **TSTSU–TO–HAGANE**
    [Overseas edition–selected articles]
    Trans. begins with 1960, v. 46+
    Order from: ISIJ

50. **THERAPIA HUNGARICA, HUNGARIAN MEDICAL JOURNAL (*)**
    Publication begins with 1953+
    Order from: TH

51. **USSR SCIENTIFIC ABSTRACTS (*)**
    Issued in various series.
    Publication begins with 1963+
    Order from: NTIS

52. **VESTNIK AKADEMII NAUK SSSR (*)**
    (Vestnik of the USSR Academy of Sciences)
    Trans. begins with 1966, v. 36+
    Monthly
    Order from: NTIS
    9.

53. **ZHURNAL NEVROPATOLOGII I PSIKHIATRII IMENI S. S. KORSAKOVA**
    (Korsakov Journal of Neurology and Psychiatry)
    Selected articles.
    Translation covers 1960, v. 60, nos. 1–3, and 1962, v. 62, nos. 1–4.
    Order from: PP
    (Soviet Neurology and Psychiatry)
    Articles selectively translated from various issues of the original.
    Order from: IASP

PART III

TRANSLATED TITLES CROSS-REFERENCED TO ORIGINAL TITLES

[Numbers in parentheses refer to citations in either Part I or Part II]

Abstracts Journal of Metallurgy (I-199)

Abstracts of Japanese Medicine (I-93)

Abstracts of Soviet Medicine (I-152)

Advances in the Physical Sciences (I-236)

Acta Chimica Sinica (I-92)

Alambre (I-56)

Algebra and Logic (I-10)

Annals of Chemistry (I-200)

Antibiotics (I-12)

Applied Biochemistry and Microbiology (I-179)

Applied Electrical Phenomena (I-61)

Applied Solar Energy (I-79)

Archives of Biological Sciences (I-15)

Archives of Immunology and Experimental Therapy (I-14)

Artificial Earth Satellites (I-98)

Astrophysics (I-17)

Atmospheric and Ocean Physics (I-105)

Automatic Control (I-25)

Automatic Documentation and Mathematical Linguistics (II-24)

Automatic Welding (I-22)

Automation and Remote Control (I-24)

Autometry (I-26)

Aviation and Cosmonautics (I-21)

Aviation Engineering (I-114)

Biochemistry (I-29)

Biological Review (I-30)

Biomedical Engineering (I-151)

Biophysics (I-28)

Bulletin of Experimental Biology and Medicine (I-31)

Bulletin of the Academy of Sciences of the USSR Physical Series (I-104)

Bulletin of the Academy of Sciences, USSR. Division of Chemical Science (I-109)

Bulletin. Academy of Sciences USSR. Geophysics Series (I-107)

Bulletin of the Boris Kidric Institute of Nuclear Science (I-27)

Bulletin of the Chemical Society, Belgrade (I-91)

Bulletin of the Institute for Geological and Geophysical Research (I-241)

Bulletin of the Moscow Society for Natural Research, Geological Section (I-33)

Cement (I-230)

Chemical and Petroleum Engineering (I-126)

Chemical Engineering (I-127)

Chemical Engineering, Japan (I-122)

Chemistry and Technology of Fuels and Oils (I-130)

Chemistry of Hetrocyclic Compounds (I-129)

Chemistry of Natural Compounds (I-131)

Chinese Journal of Physics (I-6)

Chinese Mathematics (I-3)

Clinical Medicine (I-135)

Coke and Chemistry, USSR (I-136)

[Numbers in parentheses refer to citations in either Part I or Part II]

Colloid Journal of the USSR (I-137)

Combustion, Explosion, and Shock Waves (I-68)

Computer Elements and Systems (I-256)

Contemporary Agriculture (I-203)

Corrosion Control Abstracts (I-197)

Cosmic Research (I-139)

Cybernetics (I-133)

Cybernetics Abstracts (I-196)

Cytology (I-231)

Defectoscopy: The Soviet Journal of Non-Destructive Testing (I-34)

Derwent Russian Patent Report—Chemistry and Allied Subjects (I-32)

Differential Equations (I-40)

Doklady Biochemistry (I-41)

Doklady Biological Sciences (I-41, I-42, I-43, I-44)

Doklady Biophysics (I-43)

Doklady Botonical Sciences (I-44)

Doklady Chemical Technology (I-45)

Doklady Chemistry (I-46)

Doklady of the Academy of Sciences of the USSR, Earth Science Sections (I-47, I-51, I-52)

Doklady Physical Chemistry (I-48)

Doklady Soil Science Sections (I-173)

Ecology (I-57)

Economic Geology USSR (I-84)

Electric Technology USSR (I-59)

Electrical Engineering in Japan (I-35)

Electronics and Communications in Japan (I-37, I-38)

Energie Atomique (I-20)

Engineering Cybernetics (I-112)

Engineering Journal (I-96)

Entomological Review (I-65)

Enviromental Space Sciences (I-138)

Epidemiological Review (I-188)

Experimental Medicine and Microbiology (I-153)

Exploration Geophysics (I-180)

Fibre Chemistry (I-125)

Filo Metallico, II (I-56)

Fluid Dynamics (I-102)

Friction and Wear in Machinery (I-223)

Functional Analysis and Its Applications (I-78)

Genetics (I-80)

Geochemistry (I-82)

Geochemistry International (I-82)

Geodesy and Aerophotography (I-116)

Geodesy and Cartography (I-81)

Geomagnetism and Aeronomy (I-85)

Geotectonics (I-86)

Giessereiwessen (I-143)

Glass and Ceramics (I-214)

Heat and Power (I-221)

Heliotechnology (I-79)

Herald of Communications (I-247)

[Numbers in parentheses refer to citations in either Part I or Part II]

Herald of the Academy of Medical Sciences (I-242)

High Energy Chemistry (I-132)

High Temperature (I-222)

Hydrobiological Journal (I-87)

Hydrotechnical Construction (I-88)

Hygiene and Sanitation (I-89)

Industrial Laboratory (I-261)

Industrial Wood Processing (I-39)

Information Bulletin of the Soviet Antarctic Expedition (I-207)

Inorganic Materials (I-111)

Instrument Construction (I-176)

Instruments and Experimental Techniques (I-178)

International Geology Review (I-108, I-171)

Izvestiya of the Academy of Sciences of the USSR. Geologic Series (I-108)

Izvestiya VUZOV. Instrument Building (I-117)

Izvestiya VUZOV. Radio Engineering (I-119)

Izvestiya VUZOV. Radiophysics (I-118)

Japanese Journal of Applied Physics (II-30)

JETP Letters (I-262)

Journal of Analytical Chemistry of the USSR (I-263)

Journal of Applied Chemistry of the USSR (I-271)

Journal of Applied Mathematics and Mechanics (I-181)

Journal of Applied Mechanics and Technical Physics (I-272)

Journal of Applied Spectroscopy (I-273)

Journal of Engineering Physics (I-95)

Journal of Evolutionary Biochemistry and Physiology (I-265)

Journal of General Chemistry of the USSR (I-269)

Journal of Icthiology (I-250)

Journal of Inorganic Chemistry (I-268)

Journal of Microbiology, Epidemiology, and Immunobiology (I-267)

Journal of Organic Chemistry of the USSR (I-270)

Journal of Nuclear Energy (I-20)

Journal of Physics (II-22)

Journal of Scientific Agricultural Research (I-16)

Journal of Structural Chemistry (I-274)

Journal of the Electrochemical Society of Japan (I-36)

Kinetics and Catalysis (I-134)

Korsakov Journal of Neurology and Psychiatry (II-53)

Krupp Technical Review (I-216)

Labor Hygiene and Occupational Diseases (I-90)

Lebedev Physics Institute Series (I-224)

Lithology and Mineral Resources (I-144)

Machine Science Abstracts (I-146)

Machine Tool Engineering and Production News (I-258)

Machines and Tooling (I-212)

Magnetohydrodynamics (I-145)

Mathematical Notes (I-148)

Mathematics of the USSR—Sbornik (I-147)

Measurement Techniques (I-100)

[Numbers in parentheses refer to citations in either Part I or Part II]

Mechanics of Solids (I-97)

Medical Industry of the USSR (I-149)

Medical Journal (I-142)

Medical Publishing Plan, Academy of Medical Sciences, USSR (II-3)

Medical Radiology (I-150)

Mendeleev Chemistry Journal (I-276)

Metal Science and Heat Treatment (I-155)

Metallurgist (I-156)

Meteoritica (I-157)

Meteorology and Hydrology (II-23)

Microbiology (I-158)

Microelectronics (I-159)

Military Medical Journal (I-249)

Mining and Metallurgy (I-201)

Molecular Biology (I-160)

Moscow University Chemistry Bulletin (I-245)

Moscow University Mathematics Bulletin (I-246)

Moscow University Mechanics Bulletin (I-246)

Moscow University Physics Bulletin (I-244)

NTZ-Communications Journal (II-27)

Neurophysiology (I-162)

News of Higher Educational Institutions: Instrument Building (I-117)

News of Higher Educational Institutions: Radio Engineering Series (I-119)

News of Higher Educations Institutions: Radiophysics Series (I-118)

Nonferrous Metallurgy (I-121)

Nucleonics (I-163)

Oceanology (I-165)

Optics and Spectroscopy (I-167)

Organic Reactivity (I-195)

Paleontological Journal (I-171)

Pavlov Journal of Higher Nervous Activity (I-278)

Pawlow-Zeitschrift fur Hohere Nerventatigkeit (I-278)

Petroleum Chemistry, USSR (I-161)

Petroleum Geology (I-83)

Pharmaceutical Chemistry Journal (I-128)

Pharmaceutics (I-13)

Pharmacology and Toxicology (I-66)

Physics of Metals and Metallography (I-70)

Physics of the Solid Earth (I-106)

Physiology and Biochemistry of Cultivated Plants (I-75)

Polish Chemical Industry (I-189)

Polish Endocrinology (I-64)

Polish Review of Radiology and Nuclear Medicine (I-174)

Polymer Mechanics (I-154)

Polymer Science USSR (I-257)

Probleme der Kybernetik (I-184)

Problems of Cosmogony (I-251)

Problems of Cybernetics (I-184)

Problems of Hematology and Blood Transfusion (I-183)

[Numbers in parentheses refer to citations in either Part I or Part II]

Problems of Information Transmission (I-185)

Problems of Oncology (I-253)

Problems of Psychology (I-254)

Problems of the North (I-187)

Problems of Virology (I-255)

Proceedings of the Academy of Sciences of the USSR. Agrochemistry Section (I-49)

Proceedings of the Academy of Sciences of the USSR. Applied Physics Section (I-50)

Proceedings of the Academy of Sciences of the USSR. Geochemistry Section (I-47, I-51)

Proceedings of the Academy of Sciences of the USSR. Geological Sciences Section (I-47, I-52)

Proceedings of the Academy of Sciences of the USSR. Soil Science Sections (I-53)

Proceedings of the Steklov Institute of Mathematics of the Academy of Sciences of the USSR (I-228)

Progress in Physiological Sciences (I-237)

Prospection et Protection du Sous-Sol (I-194)

Protection of Metals (I-260)

Publishing Plan, Academy of Sciences, USSR (II-4)

Questions de Micropaleontologie (I-252)

Radio Engineering (I-192)

Radio Engineering and Electronic Physics (I-193)

Radio Engineering and Electronics (I-193)

Radio Physics and Quantum Electronics (I-118)

Radiobiology (I-190)

Radiochemie (I-191)

Radiochemistry (I-191)

Refractories (I-164)

Russian Castings Production (I-143)

Russian Chemical Reviews (I-238)

Russian Engineering Journal (I-243)

Russian Journal of Inorganic Chemistry (I-268)

Russian Journal of Physical Chemistry (I-266)

Russian Mathematical Surveys (I-239)

Russian Metallurgy (I-103)

Russian Metallurgy and Fuels (I-103)

Russian Metallurgy and Mining (I-103)

Russian Patent Abstracts (I-32)

Russian Patent Bulletin—Chemistry and Fuels (I-32)

Russian Patent Gazette (I-32)

Russian Pharmacology and Toxicology (I-66)

Russian Review of Biology (I-240)

Sechenov Physiological Journal of the USSR (I-74)

Siberian Chemistry Journal (I-113)

Siberian Mathematical Journal (I-204)

Soil Mechanics and Foundation Engineering (I-169)

Solar Systems Research (I-18)

Soviet Abstracts: Mechanics (I-198)

Soviet Aeronautics (I-114)

Soviet Antarctic Expedition Information Bulletin (I-94)

Soviet Applied Mechanics (I-182)

Soviet Astronomy—AJ (I-19)

Soviet Atomic Energy (I-20)

[Numbers in parentheses refer to citations in either Part I or Part II]

Soviet Automatic Control (I-23)

Soviet Chemical Industry, The (I-124)

Soviet Electrical Engineering (I-63)

Soviet Electrochemistry (I-60)

Soviet Engineering Journal (I-96)

Soviet Fluid Dynamics (I-102)

Soviet Genetics (I-80)

Soviet Geology (I-208)

Soviet Inventions and Patents (I-170)

Soviet Inventions Illustrated (I-32)

Soviet Journal of Atomic Energy (I-20)

Soviet Journal of Developmental Biology (I-166)

Soviet Journal of Ecology (I-57)

Soviet Journal of Instrumentation and Control (I-177)

Soviet Journal of Non-Destructive Testing, The (I-34)

Soviet Journal of Nonferrous Metals (I-232)

Soviet Journal of Nuclear Physics (I-259)

Soviet Journal of Optical Technology (I-168)

Soviet Journal of Particles and Nuclei (I-67)

Soviet Journal of Quantum Electronics (I-141)

Soviet Materials Science (I-72)

Soviet Mathematics—Doklady (I-54)

Soviet Mining Science (I-73)

Soviet Neurology and Psychiatry (II-53)

Soviet Oceanography (I-210)

Soviet Physics—Acoustics (I-9)

Soviet Physics—Crystallography (I-140)

Soviet Physics—Doklady (I-55)

Soviet Physics—JETP (I-264)

Soviet Physics—Semiconductors (I-69)

Soviet Physics—Solid State (I-71)

Soviet Physics—Technical Physics (I-275)

Soviet Physics—Uspekhi (I-236)

Soviet Physics Journal (I-115)

Soviet Plant Physiology (I-76)

Soviet Plastics (I-172)

Soviet Powder Metallurgy [and Metal Ceramics] (I-175)

Soviet Power Engineering (I-58)

Soviet Progress in Chemistry (I-234)

Soviet Public Health (I-209)

Soviet Radio Engineering (I-119)

Soviet Radiochemistry (I-191)

Soviet Radiophysics (I-118)

Soviet Research in Geophysics (I-225)

Soviet Rubber Technology (I-123)

Soviet Soil Science (I-173)

Sowjetisches Gesundheitsschutz (I-209)

Sowjetisches Gesundheitwessen (I-209)

Space Biology and Medicine (I-138)

Stahl (I-211)

Stal' (in English) (I-211)

Strength of Materials (I-186)

[Numbers in parentheses refer to citations in either Part I or Part II]

Studies of Clouds, Precipitation and Thunderstorm Electricity (I-99)

Synthesen Organischer Verbindungen (I-206)

Syntheses of Heterocyclic Compounds (I-205)

Technical Cybernetics (I-112)

Technical Physics of the USSR, Leningrad (II-48)

Technology of the Textile Industry, USSR (I-120)

Telecommunications (I-62)

Telecommunications and Radio Engineering (I-62, I-192)

Theoretical and Experimental Chemistry (I-217)

Theoretical and Mathematical Physics (I-218)

Theoretical Cybernetics Abstracts (I-196)

Theoretical Foundations of Chemical Engineering (I-219)

Theory of Probability and Its Applications (I-220)

Thermal Engineering (I-221)

Transactions of the Academy of Science USSR. Mathematics Series (I-110)

Transactions of the Moscow Mathematical Society (I-229)

Trefile, Le (I-56)

Ukrainian Journal of Chemistry (I-234)

Ukrainian Mathematical Journal (I-235)

Ukrainian Physics Journal (I-233)

USSR Bulletin of Patents and Inventions (I-32)

USSR Computational Mathematics and Mathematical Physics (I-277)

Vestnik of the USSR Academy of Medical Sciences (I-242)

Vestnik of the USSR Academy of Sciences (II-52)

Water Resources (I-248)

Welding Production (I-215)

Wire: Coburg (I-56)

Wire World International (II-11)

Works of the Cerni Institute for the Development of Water Resources (I-202)

Works of the Institute of Higher Nervous Activity. Pathophysiological Series (I-227)

Works of the Institute of Higher Nervous Activity. Physiological Series (I-226)

PART IV

SOME FREQUENTLY ENCOUNTERED RUSSIAN ABBREVIATIONS OF SOVIET JOURNALS AND THEIR FULL TITLES

(Restricted to titles cited in either Part I or II of this compilation.)

ASTRON. VEST. --
  Astronomicheskii Vestnik

ASTRON. Zh. --
  Astronomicheskii Zhurnal

ATOM. ENERG. --
  Atomnaya Energiya

AVTOM. SVARKA --
  Avtomaticheskaya svarka

AVTOMET. --
  Avtometriya

BYULL. EKSPERIM. BIOL. I MED. --
  Byulleten' eksperimental'noi biologii i meditsiny

BYULL. IZ. --
  Byulleten' izobretenii

BYULL. IZ. I TOV. ZN. --
  Byulleten' izobretenii i tovarnykh

BYULL. MOSK. O-VA ISPYT. PRIRODY. OTD. GEOL. --
  Byulleten' Moskovskogo obshchestva ispytatelei prirody, otdel geologicheskii

DEREVOOBRABAT. PROM-ST' --
  Derevoobrabatyvayushchaya promyshlennost'

DAN SSSR --
  Doklady Akademii Nauk SSSR

DOKL. AN SSSR --
  Doklady Akademii Nauk SSSR

ELEKTRONNAYA OBRA. MAT. --
  Elektronnaya obrabotka materialov

ENTOMOL. OBOZRENIE --
  Entomologicheskoe obozrenie

FARMAKOL. I TOKSIKOLOGIYA --
  Farmakologiya i toksikologiya

FIZ. GOR. I VZR. --
  Fizika goreniya i vzryva

FIZ. MET. I METALLOVED. --
  Fizika metallov i metallovedenie

FIZ. TEKH. POL. --
  Fizika i tekhnika poluprovodnikov

FIZ. TV. TELA --
  Fizika tverdogo tela

FIZ. TVER. TELA --
  Fizika tverdogo tela

FIZ-KHIM. MEKH. MAT. --
  Fiziko-khimicheskaya mekhanika materialov

FIZ-TEKH. PROB. RAZ. POL. ISK. --
  Fiziko-tekhnicheskie problemy razrabotki poleznykh iskopaemykh

FIZIOLOG. Zh. SSSR IM. SECHENOVA --
  Fiziologicheskii Zhurnal SSSR imeni I. M. Sechenova

FUNKS. ANAL. EGO PRILOZH. --
  Funksional'nye analyz i ego prilozheniza

GEOD. I KARTOG. --
  Geodeziya i kartografiya

GEOL. NEFT. I GAZ. --
  Geologiya nefti i gaza

GEOL. RUDNYKH MESTOROZH. --
  Geologiya rudnykh mestorozhdenii

GEOMAG. I AERO. --
  Geomagnetizm i aeronomiya

GIDROTEKHN. STR-VO --
  Gidrotekhnicheskoe stroitel'stvo

IAN --
  Izvestiya Akademii Nauk SSSR. Seriya khimicheskaya [followed by a series title]

IFZh --
  Inzhenerno-fizicheskii zhurnal

IZh --
  Inzhenernyi zhurnal

I Zh. MEKHAN. TVER. TELA --
  Inzhenernyi zhurnal mekhanika tverdogo tela

IZV. AN SSSR, MEKHAN. ZHID. I GAZOV --
  Izvestiya Akademii Nauk SSSR. Mekhanika zhidkosti i gazov

IZV. AN SSSR, MET. --
    Izvestiya Akademii Nauk SSSR. Metally

IZV. AN SSSR, MET. I GORNOE DELO --
    Izvestiya Akademii Nauk SSSR. Metallurgiya i gornoe delo

IZV. AN SSSR, OTD. KHIM-N. --
    Izvestiya Akademii Nauk SSSR. Otdelenie khimicheskaya nauk

IZV. AN SSSR, SER. FIZ. --
    Izvestiya Akademii Nauk SSSR. Seriya fizicheskaya

IZV. AN SSSR, SER. FIZ. ATMOS. I OKEANA --
    Izvestiya Akademii Nauk SSSR. Seriya fiziki atmosfery i okeana

IZV. AN SSSR, SER. FIZ. ZEMLI --
    Izvestiya Akademii Nauk SSSR. Seriya fiziki zemli

IZV. AN SSSR, SER. GEOFIZ. --
    Izvestiya Akademii Nauk SSSR. Seriya geofizicheskaya

IZV. AN SSSR, SER. GEOL. --
    Izvestiya Akademii Nauk SSSR. Seriya geologicheskaya

IZV. AN SSSR, SER. KHIM. --
    Izvestiya Akademii Nauk SSSR. Seriya khimicheskaya

IZV. AN SSSR, SER' MATEM. --
    Izvestiya Akademii Nauk SSSR. Seriya matematicheskaya

IZV. AN SSSR, SER. NEORGAN. MATERIALY --
    Izvestiya Akademii Nauk SSSR. Seriya neorganicheskie materialy

IZV. AN SSSR, TEKH. KIB
    Izvestiya Akademii Nauk SSSR. Tekhnicheskaya kibernetika

IZV. SIB. OTDEL. AN SSSR, SER. KHIM. NAUK --
    Izvestiya Sibirskogo Otdeleniya Akademii Nauk SSSR, Seriya khimicheskikh nauk

IZV. VUZ, AVIAT. TEKH. --
    Izvestiya Vysshikh Uchebnykh Zavedenii. Aviatsionnaya tekhnika

IZV. VUZ, FIZIKA --
    Iavestiya Vysshikh Uchebnykh Zavedenii. Fizika

IZV. VUZ, GEOD. I AEROFOTOS. --
    Izvestiya Vysshikh Uchebnykh Zavedenii. Geodeziya i aerofotos'emka

IZV. VUZ, PRIB. --
    Izvestiya Vysshikh Uchebnykh Zavedenii. Priborostroenie

IZV. VUZ, RADIOFIZIKA --
    Izvestiya Vysshikh Uchebnykh Zavedenii, Radiofizika

IZV. VUZ, RADIOTEKHN. --
    Izvestiya Vysshikh Uchebnykh Zavedenii. Radiotekhnika

IZV. VUZ, TEKH. TEKS. PROMYSH. --
    Izvestiya Vysshikh Uchebnykh Zavedenii. Tekhnologiya tekstil'noi promyshlennosti

IZV. VUZ, TSVET. MET. --
    Izvestiya Vysshikh Uchebnykh Zavedenii. Tsvetnaya metallurgiya

KHIM. I NEFT. MASH. --
    Khimicheskoe i neftyanoe mashinostroenie

KHIM. GETERO. SOED. --
    Khimiya geterotsiklicheskikh soedinenii

KHIM. PRIROD. SOED. --
    Khimiya prirodnykh soedinenii

KHIM. TEKH. TOP. I MAS. --
    Khimiya i tekhnologiya topliv i masel

KHIM. VYSOKH. ENERGII --
    Khimiya Vysokikh energii

KLINICH. MEDITSINA --
    Klinicheskaya meditsina

KOLLODN. Zh. --
    Kolloidnyi zhurnal

KOSMICH. ISSLED. --
    Kosmicheskie issledovaniya

LITEINOE PROIZ-VO --
    Liteinoe proizvodstvo

LITOLOG. I POLEZ. ISKOPAEMYE --
Litologiya i poleznye iskopaemye

MATEM. ZEM. --
Matematicheskii zametki

MED. PROM-ST' --
Meditsinskaya promyshlennost' SSSR

MED. RADIOLOGIYA --
Meditsinskaya radiologiya

MED. REF. Zh. --
Meditsinskii Referativnyi Zhurnal

MED. TEKHN. --
Meditsinskaya tekhnika

METALLOVED. I TERMICHES. OBRA. METALLOV --
Metallovedenie i termicheskaya obrabotka metallov

MOLEK. BIOLOG. --
Molekulyarnaya biologiya

NAUCHNO-TEKHN. INFORM. --
Nauchno-tekhnicheskaya informatsiya

NAUCHNO-TEKHN. PROB. FIZ. GOREN. I VZRYVA --
Nauchno-tekhnicheskii problemy goreniya i vzryva

OPT. I SPEKTR. --
Optika i spektroskopiya

OPTIKO-MEKHAN. PROM-ST' --
Optiko-mekhanicheskaya promyshlennost'

OSNOV. FUNDAMEN. I MEKHAN. GRUN. --
Osnovaniya fundamenty i mekhanika gruntov

PALEONTOL. Zh. --
Paleontologicheskii Zhurnal

POROSHKOVAYA MET. --
Poroshkovaya metallurgiya

PRIB. I TEKHN. EKSPERIM. --
Pribory i tekhnika eksperimenta

PRIK. BIOKHIM. I MIKROBIOLOG. --
Prikladnaya biokhimiya i mikrobiologiya

PRIKL. MATEM. I MEKHAN. --
Prikladnaya matematika i mekhanika

PRIKL. MEKHAN. --
Prikladnaya mekhanika

PROBL. GEMATOL. I PERELIVANIYA KROVI --
Problemy gematologii i perelivaniya krovi

PROBL. KIB. --
Problemy kibernetiki

PROBL. PERED. INF. --
Problemy peredachi informatsii

PROBL. SEVERA --
Problemy severa

REAKTS. SPOSOB. ORGAN. SOEDINENII --
Reaktsionnaya-sposobnost' organicheskikh soedinenii

REF. Zh., KIB. --
Referativnyi Zhurnal. Kibernetika

REF. Zh., MEKHAN. --
Referativnyi Zhurnal. Mekhanika

REF. Zh., MET. --
Referativnyi Zhurnal. Metallurgiya

SIBIR. MATEM. Zh. --
Sibirskii Matematicheskii Zhurnal

SOV. GEOLOGIYA --
Sovetskaya geologiya

SOV. ZDRAVOOKHRANENIE --
Sovetskoe zdravookhranenie

TEOR. I EKSPERIM. KHIM. --
Teoreticheskaya i eksperimental'naya khimiya

TEOR. OSNOVY KHIM. TEKH. --
Teoreticheskie osnovy khimicheskoe tekhnologii

TEO. VEROY. I EE PRIMEN. --
Teoriya veroyatnostei i ee primeneniya

TEPLOFIZ. VYSOK. TEMP. --
Teplofizika vysokikh temperatur

TR. FIZ. IN-TA AN SSSR --
  Trudy Fizicheskogo Instituta imeni
  P. N. Lebedeva Akademii Nauk SSSR

TR. GEOFIZ. IN-TA AN SSSR --
  Trudy Geofizicheskogo Instituta
  Akademii Nauk SSSR

TR. MORSK. GIDROFIZ. IN-TA. AN SSSR --
  Trudy Morskogo Gidrofizicheskogo
  Instituta Akademii Nauk SSSR

TR. MOSK. MATEM. O-VA --
  Trudy Moskovskogo Matematicheskogo
  Obshchestva

TVSET. MET. --
  Tsvetnye metally

UFN --
  Uspekhi fizicheskikh nauk

UKR. KHIM. Zh. --
  Ukrainskii khimicheskii zhurnal

UMN --
  Uspekhi matematicheskikh nauk

USB --
  Uspekhi sovremennoi biologii

USPEKHI FIZ. NAUK --
  Uspekhi fizicheskikh nauk

USPEKHI MATEM. NAUK --
  Uspekhi matematicheskikh nauk

USPEKHI SOVREM. BIOL. --
  Uspekhi sovremennoi biologii

VESTN. AN MED. SSSR --
  Vestnik Akademii Meditsinskikh Nauk
  SSSR

VESTN. AN SSSR --
  Vestnik Akademii Nauk SSSR

VESTN. MASHIN. --
  Vestnik Mashinostroeniya

VESTN. MOSK. (GOS.) UNIV., FIZ., ASTRON. --
  Vestnik Moskovskogo (gosudarstvenogo)
  universiteta. Seriya fizika, astronomiya

VESTN. MOSK. (GOS.) UNIV., KHIM. --
  Vestnik Moskovskogo (gosudarstvenogo)
  universiteta. Seriya khimiya

VESTB. SVYAZI --
  Vestnik svyazi

VOENNO-MED. Zh. --
  Voenno-meditsinskii Zhurnal

VOPR. KOSMOG. --
  Voprosy kosmogonii

VOPR. ONKOLOG. --
  Voprosy onkologii

VOPR. VIRUS.
  Voprosy virusologii

VYSOKOM. SOED. --
  Vysokomolekulyarnye soedineniya

YADER. FIZ. --
  Yadernaya fizika

ZAVOD. LABOR. --
  Zavodskaya laboratoriya

Zh. ANALIT. KHIM. --
  Zhurnal Analiticheskoi khimii

Zh. EKSPERIM. I TEOR. FIZ. --
  Zhurnal eksperimental'noi i teoreticheskoi
  fiziki

ZhETF --
  Zhurnal eksperimental'noi i teoreticheskoi
  fiziki

ZhETF PIS'MA V REDAK. --
  ZhETF Pis'ma v redaktsiyu

Zh. FIZ. KHIM. --
  Zhurnal fizicheskoi khimii

ZhFKh. --
  Zhurnal fizicheskoi khimii

Zh. NEORG. KHIM. --
  Zhurnal neorganicheskoi khimii

Zh. OBSHCH. KHIM. --
  Zhurnal obshchei khimii

Zh. ORG. KHIM. --
    Zhurnal organicheskoi khimii

ZhPKh. --
    Zhurnal prikladnoi khimii

Zh. PRIK. KHIM. --
    Zhurnal prikladnoi khimii

Zh. PRIK. MEKHAN. TEKHN. FIZ. --
    Zhurnal prikladnoi mekhaniki i teknicheskoi fiziki

Zh. PRIK. SPEK. --
    Zhurnal prikladnoi spektroskopii

ZhPS --
    Zhurnal prikladnoi spektroskopii

Zh. STRUKT. KHIM. --
    Zhurnal strukturnoi khimii

Zh. TEKH. FIZ. --
    Zhurnal tekhnicheskoi fiziki

ZhTF --
    Zhurnal tekhnicheskoi fiziki

Zh. VSES. KHIM. OBSHCH. IM. MENDELEEVA --
    Zhurnal vsesoyuznogo khimicheskogo obshchestva imeni D. I. Mendeleeva

Zh. VYCH. MATEM. I MATEM. FIZ. --
    Zhurnal vychislitel'noi matematiki i matematicheskoi fiziki

Zh. VYS. NERV. DEYAT. IM. PAVLOVA --
    Zhurnal vysshei nervnoi deyatel'nosti imeni I. P. Pavlova

PART V

KEY TO
ADDRESSES OF PUBLISHERS
AND DISTRIBUTORS
ABBREVIATED IN INDIVIDUAL CITATIONS

AFS
American Fisheries Society
1319 18th Street, N.W.
Washington, D.C., USA 20036

AGI
American Geological Institute
2201 M Street, N.W.
Washington, D.C., USA 20036

AGS
American Geographical Society
Broadway at 156th Street
New York, N.Y., USA 10032

AGU
American Geophysical Union
1707 L Street, N.W.
Washington, D.C., USA 20036

AIAA
American Institute of Aeronautics
and Astronautics
1290 Avenue of the Americas
New York, N.Y., USA 10019

AINA
Arctic Institute of North America
1619 New Hampshire Avenue, N.W.
Washington, D.C., USA 20009

AIP
American Institute of Physics
335 East 45th Street
New York, N.Y., USA 10017

AMFETEX
AMFETEX Inc.
P.O. Box 213
Arlington, Mass., USA 02174

AMS
American Mathematical Society
P.O. Box 6248
Providence, Rhode Island, USA 02904

API
Academic Press, Inc.
111 Fifth Avenue
New York, N.Y., USA 10003

APr
Allerton Press, Inc.
150 Fifth Avenue
New York, N.Y., USA 10011

ARSRCDS
Academia Republici Socialiste Romania
Centrul de Documentare Stiintifica
Str. Gutenberg, nr. 3 bis
Bucuresti, Romania

ARTIA
"Artia"
Smecky 30
Praha 2, Czechoslovakia

ASCE
American Society of Civil Engineers
345 East 47 Street
New York, N.Y., USA 10017

ASME
American Society of Mechanical Engineers
345 East 47th Street
New York, N.Y., USA 10017

BCIRA
British Cast Iron Research Association
Bordesley Hall, Alvechurch
Birmingham, England

BISI
British Iron and Steel Industry
Translation Service
Iron and Steel Institute
4, Grosvenor Gardens
London, S.W. 1, England

BRGGM
Bureau de Recherches Geologiques
Geophysiques et Minieres
74, Rue de la Federation
Paris 15, France

BWRA
British Welding Research Association
Abington Hall, Abington
Cambridge, England

NOTE: "OP" following the publisher's initials in Part 1 or 2 indicates that back issues are no longer available from the publisher.

CIINTE
Centralny Instytut Informacji
   Naukowo-Technicznej i Ekonomicznej
Aleja Niepodleglosci 188
Warszawa 12, Poland

CNRS
Centre Nationale de la Recherche
   Scientifique Services des Publications
15, Quai Anatole France
Paris VIIe, France

CS
Chemical Society
Burlington House, Piccadilly
London, W. 1, England

CTRA
Coal Tar Research Association
Oxford Road
Gomersal, Leeds, England

CTT
Columbia Technical Translations
5 Vermont Avenue
White Plains, N. Y., USA 10606

DBEI
Deutscher Buch-Export und Import, GmbH
Postschliessfach 276
Leipzig C. 1, East Germany

DFC
Draht Fachzeitschrift
Bahnhofstrasse 31
8630 Coburg, Germany

DIS
Derwent Information Service
Rochdale House, Theobalds Road
London, W.C. 1, England

ECJ
Electrochemical Society of Japan
3, 1-chome, Yurakucho
Chiyoda-ku, Tokyo, Japan

ECL
Electrical Communications Laboratory (nippon)
Nippon Telephone and Telegraph,
   Public Corporation
Tokyo, Japan

EGP
Economic Geology Publishing Company
Urbana, Illinois, USA 61801

EL
Elsevier Publishing Company
P.O. Box 211
Amsterdam, The Netherlands

EMF
Excerpta Medica Foundation
2 East 103rd Street
New York, N.Y., USA 10029

ESA
Entomological Society of America
4603 Calvert Road
College Park, Maryland, USA 20740

ETP
Eagle Technical Publications
62-b Stapleton Hall Road
London, N. 4, England

EUROMED
Euromed Publications
97 Moore Park Road
London, S.W. 6, England

FASEB
Federation of American Societies
   for Experimental Biology
9650 Rockville Pike
Bethesda, Maryland, USA 20014

FP
Faraday Press, Inc.
84 Fifth Avenue
New York, N.Y. USA   10011

NOTE: "OP" following the publisher's initials in Part 1 or 2 indicates that back issues are no longer available from the publisher.

FTD
Foreign Technology Division
Air Force Systems Command
Wright-Patterson Air Force Base
Ohio

GB
Gordon and Breach, Science Pub., Inc.
150 Fifth Avenue
New York, N.Y., USA 10011

HTCBN
Hungarian Trading Company for Books and
Newspapers "Kultura"
P.O. Box 149
Budapest 62, Hungary

IASP
International Arts and Sciences Press
901 North Broadway
White Plains, N.Y., USA 10603

ICE
International Chemical Engineering
345 East 47th Street
New York, N.Y., USA 10017

IEEE
Institute of Electrical and Electronics Engineers
275 Madison Avenue
New York, N.Y., USA 10016

IPI
International Physical Index, Inc.
1909 Park Avenue
New York, N.Y., USA 10016

IPST
Israel Program for Scientific Translations, Ltd.
Kiriat Moshe
Jerusalem, Israel

ISA
Instrument Society of America
530 William Penn Place
Pittsburgh, Pa., USA 15219

ISIJ
Iron and Steel Institute of Japan
Japan Travel Bureau Bldg.
No. 1, Marunochi 1-chome, Chiyoda-ku
Tokyo, Japan

JAERI
Japan Atomic Energy Research Institute
Takai-mura, Naka-gun,
Ibaraki-ken, Japan

KP
Kraus Periodicals
16 East 46th Street
New York, N.Y., USA 10017

KRCL
Kraus Reprint Corporation
Vaduz
Liechtenstein

KRUPP
Fried. Krupp AG
Essen
West Germany

LAL
Laboratoire d'Astronomie de Lille
1, Impasse d' Obervatoire
Lille, France

MACM
Macmillan and Company, Ltd.
10-15 St. Martin's Street
London, W.C. 2, England

Ma&S
Maclaren & Sons Ltd.
P.O. Box 109
Davis House
69-77 High St.
Croyden, Surrey CR 9 1QH
England

MAR
Maruzen Co., Ltd., Export Dept.
P.O. Box 605, Tokyo Central
Tokyo, Japan

NOTE: "OP" following the publisher's initials in Part 1 or 2 indicates that back issues are no longer available from the publisher.

McE
Ralph McElroy Co., Inc.
504 West 24th St.
Austin, Texas, USA 78705

MTRC
Ministry of Technology
Reports Centre
Station Square House
St. Mary Cray
Orpington, Kent BR5 3RE
England

NFSAIS
National Federation of Science Abstracting and
  Indexing Services
324 East Capitol Street
Washington, D.C., USA 20003

NLL
National Lending Library for Science and
  Technology
Boston Spa, Yorkshire, England

NCR
National Research Council
The Library
Ottawa 2, Canada

NTIS
National Technical Information Service
5285 Port Royal Road
Springfield, Va., USA 22151

NTZ
Nachrichtentechnische Zeitschrift
14, Graf Adolf Platz
4 Dusseldorf 1, West Germany

OSA
Optical Society of America, Inc.
American Institute of Physics
335 East 45th Street
New York, N.Y., USA 10017

PERA
Production Engineering Research Assoc.
Melton Mowbray
Leicestershire, England

PG
Petroleum Geology
Box 171
McLean, Virginia, USA 22101

PLPC
Plenum Publishing Corporation
227 West 17th Street
New York, N.Y., USA 10011

PP
Pergamon Press, Inc.
44-01 21st Street
Long Island City, N.Y., USA 11101

PPC
Palmerton Publishing Company, Inc.
101 West 31st Street
New York, N.Y., USA 10001

PS
Primary Sources
11 Bleecker Street
New York, N.Y., USA 10012

P-V
Porta-Verlag
Munchen, West Germany

RAPRA
Rubber & Plastics Research Assoc.
  of Great Britain
Shawbury, Shrewsbury SY4 4NR England

RIS
Research Information Service
44-01 21st Street
Long Island City, N.Y., USA 11101

RTP
Rubber and Technical Press, Ltd.
25 Lloyd Baker Street
London, W.C. 1, England

SIAM
SIAM Publications
Box 7541
Philadelphia, Pa., USA 19100

NOTE: "OP" following the publisher's initials in Part 1 or 2 indicates that back issues are no longer available from the publisher.

SIC
Scientific Information Consultants, Ltd.
661 Finchley Road
London, N.W. 2, England

SP
Serapeion Press, Inc.
35 East 12th Street
New York, N.Y., 10003

SSSA
Soil Science Society of America, Inc.
Program Director, Soviet Soil Science
677 South Segoe Road
Madison, Wisconsin, USA 53711

ST
Scripta Technica, Inc.
1511 K Street, N.W.
Washington, D.C., USA 20005

STE
Scripta Technica (England)
138 New Bond Street
London, W.1, England

STR
Spectrum Translation and Research, Inc.
207 East 37th Street
New York, N.Y., USA 10016

TF
Taylor and Francis, Ltd.
Red Lion Court
Fleet Street
London, E.C. 4, England

TH
Therapia Hungarica
P.O. Box 64
Budapest, Hungary

TI
Textile Institute
10, Blackfriars Street
Manchester 3, England

TIL
Technical Information Library
Ministry of Aviation
First Avenue House
High Holborn
London, W.C. 1, England

TNTID
Tsentur za Nauchno-Tekhnicheska
Informatsii i Dokumentatsiia
Ul. 7 noemvri No. 1
Sofia, Bulgaria

UT
University of Tokyo
Faculty of Science
Publications Office
No. 342 Physics Building
Bunkyo-ku
Tokyo, Japan

WERK
Werk Publishing Company
Box 210
8401 Winterthur, Switzerland

WWI
Wire World International
Jahnstrasse 36
D-4 Dusseldorf
West Germany

NOTE: "OP" following the publisher's initials in Part 1 or 2 indicates that back issues are no longer available from the publisher

PART VI

RUSSIAN ALPHABET AND TRANSLITERATION
(USED IN THIS STUDY)

RUSSIAN ALPHABET AND TRANSLITERATION

English trans- literation	Russian "Roman" letter	Russian "Italic" letter	Russian script letter
a	А а	*А а*	*А а*
b	Б б	*Б б*	*Б б*
v	В в	*В в*	*В в*
g, v	Г г	*Г г*	*Г г*
d	Д д	*Д д*	*Д д*
ye, e	Е е	*Е е*	*Е е*
yë, ë	Ё ё	*Ё ё*	*Ё ё*
zh	Ж ж	*Ж ж*	*Ж ж*
z	З з	*З з*	*З з*
i	И и	*И и*	*И и*
y	Й й	*Й й*	*Й й*
k	К к	*К к*	*К к*
l	Л л	*Л л*	*Л л*
m	М м	*М м*	*М м*
n	Н н	*Н н*	*Н н*
o	О о	*О о*	*О о*
p	П п	*П п*	*П п*
r	Р р	*Р р*	*Р р*
s	С с	*С с*	*С с*
t	Т т	*Т т*	*Т т*
u	У у	*У у*	*У у*
f	Ф ф	*Ф ф*	*Ф ф*
kh	Х х	*Х х*	*Х х*
ts	Ц ц	*Ц ц*	*Ц ц*
ch	Ч ч	*Ч ч*	*Ч ч*
sh	Ш ш	*Ш ш*	*Ш ш*
shch	Щ щ	*Щ щ*	*Щ щ*
"	Ъ ъ	*Ъ ъ*	*ъ*
y	Ы ы	*Ы ы*	*ы*
'	Ь ь	*Ь ь*	*ь*
e	Э э	*Э э*	*Э э*
yu	Ю ю	*Ю ю*	*Ю ю*
ya	Я я	*Я я*	*Я я*